THE FIRST
COPERNICAN

Coat of arms of Rheticus's family on his mother's side, de Porris, which he Germanized to von Lauchen ("of the leeks").

THE FIRST
COPERNICAN

GEORG JOACHIM RHETICUS
AND THE RISE OF
THE COPERNICAN REVOLUTION

DENNIS DANIELSON

WALKER & COMPANY
NEW YORK

Published by Walker & Company, New York
Distributed to the trade by Holtzbrinck Publishers

All papers used by Walker & Company are natural, recyclable products made from wood
grown in well-managed forests. The manufacturing processes conform to the environmental
regulations of the country of origin.

Art credits: Frontispiece (Rheticus Coat of Arms), "scratchboard" etching by Kirsten Behee;
used by permission. Page 6, photograph courtesy of Uppsala University Library; used by
permission. Pages 32 and 221, photographs by Reiner Dienlin; used by kind permission.
Pages 71, 83, and 200, photographs courtesy of Linda Hall Library of Science, Engineering
& Technology, Kansas City; used by permission. Pages 78, 94, 175, and 197, photographs by
Dennis Danielson; used by kind permission of the Stadtbibliothek Feldkirch. Pages 110 and
111, photograph by Owen Gingerich; used by kind permission. Page 203, photograph by
Dennis Danielson; used by kind permission of the Stadtarchiv Lindau. Page 205,
photograph courtesy of Jagiellonian Library, Krakow; used by permission. Page 223,
photograph by Dennis Danielson.

Library of Congress Cataloging-in-Publication Data has been applied for.

ISBN-10: 0-8027-1530-3
ISBN-13: 978-0-8027-1530-2

Visit Walker & Company's Web site at www.walkerbooks.com

First U.S. edition 2006

1   3   5   7   9   10   8   6   4   2

Typeset by Westchester Book Group
Printed in the United States of America by Quebecor World Fairfield

for Janet
Sole partner, and sole part, of all these joys

CONTENTS

PROLOGUE

No Rheticus, No Copernicus

The seventy-year-old man stretched out on the disheveled surface of the bed already looked half dead—literally. Pressure from a cerebral hemorrhage had caused right-lateral paralysis of his body, so that the eyelid and cheek on that side of his face sagged lifeless. And the eyes that had gazed so deeply into the beauty of the universe no longer seemed to those present to see across the room, nor to discern the movements of people who entered it, let alone the time of day or the procession of the seasons. Like many in their last days, Nicolaus Copernicus left his attendants questioning whether his mind and spirit were still there.

Georg Donner, the other old man in the room, had been tending his patient since late autumn of the previous year when Copernicus had suffered a stroke. Their close friend and former fellow canon Tiedemann Giese, bishop of Kulm, residing seventy miles away in Löbau, had heard the bad news and on December 8, 1542, dashed off a letter to Donner in Frauenburg entreating him to look after the man he called "our revered old Copernicus." "When he was in sound health," Giese wrote, "he loved spending time alone, and for this reason I fear he may have but few friends who will stand by him now in his illness—though we all

1

owe so much to his upright character and outstanding instruction . . . Please, then, as his condition dictates, watch over him, and care for this man whom you and I together have always loved."

More than five months after that letter, Donner was still faithfully watching and caring. As he would report to Giese, to all appearances Copernicus's "memory and mental powers had abandoned him" already. Nevertheless, on this Thursday in the late spring of 1543 Georg Donner still hoped, while a little life remained, to reach his dying friend with some gesture of closure or consolation.

To those who had known Copernicus, he was many things: a professional administrator, a sought-after economist, a paragon of humanist learning, and above all a skilled doctor. But almost nobody had an inkling of how history would remember him: as the preeminent astronomer who, in the words of one Wittenberg theologian, "moved the earth and immobilized the sun."[1] Most of those acquainted with Copernicus had only sketchy notions of his deep commitment to the science of the heavens and an even dimmer grasp, if any, of his radiant cosmological idea.

This was hardly surprising. Copernicus was technically an amateur astronomer. Unlike his most famous successors—Tycho Brahe, Johannes Kepler, Galileo Galilei, Isaac Newton—Copernicus was never paid for his science nor ever granted any patronage to help him pursue it. Not once since his student days in Italy had he ever delivered a lecture on celestial mechanics or mathematics in any academy or university. Some thirty years before, in the second decade of the sixteenth century, he had circulated a manuscript containing his radical core ideas about a moving earth and a stationary sun, but nobody seemed to respond or to pay him the slightest attention.

That was still the state of things when four years earlier twenty-five-year-old Georg Joachim Rheticus became aware of the theories of the elderly amateur astronomer and felt compelled to investigate them firsthand. A mathematics professor from Martin Luther's University of Wittenberg, Rheticus—whose own father, a doctor-scientist, had been beheaded as a swindler and thief eleven years earlier—chanced the long pilgrimage to Varmia in hopes of meeting Copernicus and ended up staying with him for two and a half years. His arrival in late May of 1539 threw both men into a state of astonishment: Rheticus at hearing such world-bending ideas straight from the man whom he would eventually "revere as a father,"[2] and Copernicus at acquiring a real student, the only one he would ever have.

When Rheticus arrived in 1539, Georg Donner was not yet, like Copernicus, one of the sixteen Frauenburg canons who administered the cathedral and, as the bishop's deputies, ran the province of Varmia. But he already knew Copernicus well enough to be aware of his devotion to the heavens and to long hours of private study, and he may have heard about a letter Copernicus had received from a cardinal in Italy urging him to expound on his "new cosmology,"[3] though such expressions of interest were rare. Especially in Varmia and Prussia, not renowned for their intellectual ferment, anyone hearing rumors of the old canon's incredible theory—that the earth was twirling about in the heavens—would simply have ignored it or laughed it off as a product of creeping senility.

Rheticus, however, had the imagination, the mathematics, and the desire to think his way into Copernicus's new world. Indeed for him the experience was much like an infatuation—except that the beautiful face that obsessed him, held him captive, absorbed his mind and his waking hours, belonged not to a person but to the cosmos.

As a result, Rheticus grew exhausted and pale in those early summer months of 1539.[4] Yet by autumn he had rallied and quickly began hatching grand schemes for heralding his teacher's vision of beauty. In early 1540 he launched a Copernican trial balloon—publishing in nearby Danzig a brilliant précis or *First Account* of the new cosmology replete with respectful language about honoring one's predecessors. Then the intrepid Rheticus wasted no time sending copies of the small

volume out to key readers, not only to the Catholic bishop in Heilsberg and the Protestant duke in Königsberg, but also farther afield: to
a famous cartographer in the Low Countries, to his closest astrologer
friend away to the southwest on the borders of Switzerland, and, most
decisively, to one of Germany's leading scientific publishers, Johannes
Petreius of Nuremberg.

Not that anybody actually believed what Rheticus presented of his
teacher's cosmological ideas; yet firm written evidence of serious curiosity on the part of scholars around Europe was enough. Together
with continuous encouragement by Rheticus and Giese and similarly
documented eagerness on the part of the Nuremberg publisher, that serious curiosity tipped the scales: Copernicus at last agreed to share his
work with the world.

This meant finishing it as well as transcribing a manuscript that his
only disciple could deliver to the printing house. But all went according to Rheticus's plans; by early autumn of 1541, when he was ready to
leave Frauenburg and bid a final farewell to his beloved teacher, the astronomical life's work of Copernicus was carefully packed for the long
journey south.

Rheticus's apostolic mission to Nuremberg then encountered a series of difficulties. In December 1542, when the laborious publication
process was still far from complete, Copernicus suffered his stroke.
Moreover, by this time Rheticus was beginning a new academic position in Leipzig and unfortunately had to leave supervision of the printing to someone else.

Finally, in April of 1543, the great work emerged from the presses
and was made ready for delivery. In keeping with sixteenth-century
custom, the book's broad printed sheets, each of which would be
folded once to form two leaves (four "folio" pages), were shipped loose,
a practice allowing purchasers to have the work custom bound by their
own local bookbinder.

By early May Donner knew of the arrival of the parcel from

Nuremberg, though at first he may have thought little of it, given Copernicus's low state. But on May 24, remembering Rheticus and the exertion and exhilaration of that unlikely paternal-filial enterprise, Donner decided to perform a remarkable gesture.

He sent for the sheaf of pages that embodied Copernicus's masterpiece. At the top of the rather spare title page the author's name appeared in block capitals, NICOLAUS COPERNICUS OF TORUŃ, followed in smaller type by the work's subject and title: *On the Revolutions of the Heavenly Spheres*. This book, Donner sensed, would be Copernicus's chief legacy and monument. For himself, he would cherish his own personal copy, presented and signed by Rheticus "to my friend the Reverend Mr. Georg Donner, Canon of Varmia." Although he had every reason to doubt whether the book's semiparalysed and gravely ill author could any longer comprehend his surroundings, Donner carefully placed the recently arrived work in front of Copernicus so that he could see it.* And only a short while after those fresh pages were placed upon his bed, Copernicus took a labored last breath, and was silent.[5]

KRAKÓW, SOUTHERN POLAND,
MONDAY, OCTOBER 28, 2002

Sometimes when people travel alone, they talk to themselves. "Well, what did you really come here for?" I asked myself aloud as I swung out of bed one bleak autumn morning in Kraków, where I had arrived late the night before.

It was Monday, and I had forty-eight hours to experience this beautiful former royal capital of Poland. While the following day would be taken up with academic meetings and a lecture on Copernicus I was to give at the Jagiellonian University, today was the only unplanned stretch of time in my entire Polish pilgrimage. Part of me wanted to

---

*Donner would later tell Giese, who repeated it to Rheticus, that Copernicus did see it.

NICOLAI CO

PERNICI TORINENSIS

DE REVOLVTIONIBVS ORBI=
um cœlestium, Libri VI.

Habes in hoc opere iam recens nato, & ædito,
studiose lector, Motus stellarum, tam fixarum,
quàm erraticarum, cum ex ueteribus, tum etiam
ex recentibus obseruationibus restitutos: & no=
uis insuper ac admirabilibus hypothesibus or=
natos. Habes etiam Tabulas expeditissimas, ex
quibus eosdem ad quoduis tempus quàm facilli
me calculare poteris. Igitur eme, lege, fruere.

Ἀγεωμέτρητος οὐδεὶς εἰσίτω.

Collegii Brunsbergensis Societati Jesu.

Norimbergæ apud Ioh. Petreium,
Anno M. D. XLIII.

Reverendo D. Georgio
Donner canonico Varmiensi
amico suo ...
... Rheticus F.J.

*What Copernicus saw (except for the handwriting, which includes Rheticus's
dedication to Donner—crossed out most likely by one of the volume's
subsequent stewards, a Jesuit librarian).*

visit the famous salt mines in Wieliczka, but the weather was terrible. Still, despite the wind and cold, I had a day to do anything I wanted. I could not bear the thought of wasting it.

I had come to Poland to trace the footsteps of Copernicus and had already visited his birthplace, Toruń, lecturing at the university that now bears his name. I had also toured his later places of residence in the northern province of Varmia: Lidzbark (Heilsberg), Olsztyn (Allenstein), and in particular Frombork (Frauenburg). And for me it had been a rewarding journey indeed. But continuously slipping in and out of the shadows along the trail of Copernicus was a second, less distinct, yet even more beguiling figure, whose name was Georg Joachim Rheticus.

Every pamphlet writer or guide sketching the story of Copernicus would mention Rheticus, the great astronomer's "only student" or "sole apostle," who sought out Copernicus in his old age, convinced him to publish his *Revolutions,* and then saw to it that the publication actually happened. Right from Tiedemann Giese's acknowledgment in 1543 that Rheticus was "chief impresario" of this drama, history attested the indispensable role he had played: no Rheticus, no Copernicus.[6] But nobody seemed to explain what spurred Rheticus to do all that, or how he did it, or even who he was, or what became of him after his teacher died with Rheticus not yet thirty, or exactly how his subsequent trigonometric research connected with his heroic mission as the first Copernican.

So on that windy Monday morning in Kraków, when I wondered aloud what I really wanted to do, the answer shot from within me instantly and without reflection: "I want to meet Rheticus."

I already knew that, after fleeing a scandal in Leipzig, Rheticus spent most of the last twenty years of his life in Kraków, more than half a century after Copernicus himself attended the city's renowned Jagiellonian University as a young student. His framed matriculation record—"Nicolaus Nicolai" of Toruń—still hangs in the museum of the university's Collegium Maius, along with a handwritten tribute by the astronaut Neil Armstrong penned in 1973 on the occasion of Copernicus's five-hundredth birthday. But I could see no such monuments to Rheticus. If I hoped ever to meet him, I wondered where I should look.

I remembered I had brought along a letter of introduction from my own institution, the University of British Columbia, requesting that I be granted reading privileges at the Jagiellonian Library. Furthermore, I knew that the Jagiellonian owned rare copies of two of Rheticus's landmark works on trigonometry. I hesitated. Part of me doubted whether only one day and about five words of Polish left me any chance of gaining access to these treasures.

Still, I bundled up and ventured out on foot. With the help of some students on their way to classes, I found the library and entered its grand foyer. There stood a large plaque listing departments—in Polish, of course—and a gloomy-looking guard who simply shook his head when I addressed him in English, then German, then French. I recalled that the sign over my hotel's check-in counter looked a lot like RECEPTION, so I slowly pronounced this word in a way I hoped sounded vaguely Polish. The guard's face brightened somewhat, and he likewise replied with a single word, "Secretariat"—and gestured to his right, down the corridor. I muttered "Dziękuję" and hurried off to inspect a variety of closed, windowless doors, one of which bore the sign SEKRETARIAT.

I swallowed hard and did as the Europeans do: knocked lustily and, without waiting for a response, opened the door, took a step into the room, and awaited my fate. One kind woman, hearing English sounds drop from my lips and perhaps recognizing the helpless-foreign-scholar look on my face, headed off into the passageway to find someone to whom I could explain my quest.

A patient English-speaking librarian soon arrived and helped me obtain a reader's card. She then passed me over to a rare-books librarian,

Grażyna Stępień, who took me to a cryptlike room housing the old catalog of books, with titles and call numbers ("signatures") appearing on handwritten slips of paper arranged alphabetically in small, shallow wooden drawers. After some confusion caused by the variable spellings of the name Rheticus, we found what we were looking for, and Ms. Stępień helped me fill out the order slips.

Successfully placing a book order in Polish was itself a small victory, though I knew that in many great libraries one might wait hours before books are actually delivered. But Ms. Stępień led me straight to the rare-books reading room, sat me down at an ordinary library table, and disappeared. There I waited uncomfortably, cold and damp with nervous sweat, listening to the windows rattling from the storm that had not yet blown itself out. Within ten minutes Ms. Stępień returned triumphantly with two books, one huge leather-bound folio, and one thin quarto-format pamphlet. I thanked her—and shivered.

I gently opened the smaller volume to its title page: *Canon doctrinae triangulorum (Canon of the Science of Triangles)*. There was Rheticus's trademark obelisk symbolizing ancient astronomical observation. This twenty-four-page pamphlet printed in Leipzig in 1551 contained the first six-function trig tables ever published, followed by a dialogue between a semifictional teacher (Philomathes—"math lover" or "lover of learning," an alter ego of Rheticus himself) and an inquiring stranger (Hospes). Philomathes proclaimed this new science of triangles to be both useful and pleasant. He announced, moreover, that the *Canon*'s author was a man delivering "fruits from the most delightful gardens of Copernicus."

I closed the thin volume and reached with both hands for the other one, a book that on the outside resembled a large dictionary. It comprised about fifteen hundred rather stiff folio pages mainly filled with expanded six-function trig tables, typically showing values to ten decimal places for angles in increments of ten seconds (one 360th of a degree)—except that in the sixteenth century decimals were not yet in use, so each entry was written as a ratio over 10,000,000,000.

I knew that this astoundingly labor-intensive work, whose calculations were all computed by human beings, not machines, was the very fountainhead of what would later be called trigonometry. The volume bore the title *Opus palatinum* in honor of the Palatinate prince in Heidelberg who was its biggest financial backer. I also knew that Rheticus died in 1574 and left his work in the hands of his student Valentin Otto. Because Otto took more than two decades to finish it, the book did not appear until 1596. But this heavy volume was the fruit of Rheticus's mathematical and Copernican labors, and I hoped that here I might catch a closer glimpse of the mathematician himself.

I turned back to the beginning of the *Opus*. On the title page stood two obelisks, perhaps because the work had two authors. Then, in the opening section, instead of tables I found a long prefatory letter addressed to Otto's "illustrious reader." Having read that at some point in this preface Otto spoke in personal terms about his relationship with his teacher, I urgently scanned these large pages hoping for a glimmer of revelation.

And through the thick fog of Otto's Latin prose a scene began to take shape. Otto abruptly digressed from his mathematical introduction to recount how, while he was still in his early twenties, his reading of the dialogue in Rheticus's 1551 *Canon* "aroused and inflamed" him with a desire to leave Wittenberg and embark on an odyssey in search of Rheticus some hundreds of miles away in Hungary. He told of the warm welcome he received, and of how Rheticus instantly saw in him a reflection of himself as he had been when in his youth he visited Copernicus on his own quest for news of a new science.

I read of the immediate and growing collaboration of this student-teacher pair, then of Rheticus's suddenly declining health, and of how he bequeathed his life's work to his student—and how finally, but so soon, at the age of sixty, as Otto wrote, "at about two o'clock during the night, the most beloved teacher Georg Joachim Rheticus expired in my arms."

Absorbed by Otto's account, I began to comprehend how the heavy book I held in my hands—a descendant of the labors of Copernicus and the foundational work for subsequent centuries of trigonometric practice in geography, surveying, architecture, and astronomy—had its beginning in two men's passion for triangles, thirst for knowledge, and affection for their teachers and for each other. And though the wind continued to whistle in the streets of Kraków, those pages enveloped me for a time in an ineffable quiet reverence. For a few moments, at least, more than four hundred years after his death, I had met Rheticus.

A NOTE ON NAMES

Telling stories, including stories about science or scientists, involves acts of translation. And a translator's choice of words, given the twists and turns of history, can be a delicate matter, especially if the words are the names of people or places.

In this account of the life of Rheticus my choices have been guided by pragmatism and a desire to communicate accurately, with no offense intended to anyone who might prefer words different from the ones I use. In some cases, my usage is determined by English language convention (Nuremberg rather than Nürnberg); in others, by precedence (Frauenburg rather than Frombork, the latter being a Polonization of the former); in still others, by diplomatic compromise among competing candidates. Hence I write Cassovia (Latin), commonly used in the sixteenth century, rather than Kaschau (German), Kassa (Hungarian), or Košice (Slovak), by which that city is rightly known today.

As for personal names, there was huge variation among these in the sixteenth century; in particular, a large proportion of scholars Latinized their names. In each case I attempt a reasonable choice within the range of what characters were actually called by their contemporaries. For the sake of simplicity I use Otto rather than Otho, and of course Rheticus rather than other spellings sometimes found in different contexts and languages (Rhetikus in German, Retyk in Polish, and Rhaeticus or Rhäticus in various documents and library catalogs).

Although many scholars prefer original Latin book titles, I have opted for straightforward English equivalents—*The Revolutions* for *De revolutionibus* (Copernicus); *First Account* for *Narratio prima* (Rheticus), and so on—while occasionally indicating or retaining a Latin title for the sake of clarity (e.g., *Opus palatinum*). Behind all of these choices is the hope that the vigor of Renaissance language and the richness of the career and milieu of Rheticus might resonate with fidelity in modern English.

CHAPTER 1

PATRONS AND A POET

Georg Joachim Rheticus was born in the alpine town of Feldkirch on February 16, 1514. Feldkirch stood on a major trade route linking Italy and Germany, and Rheticus's family and education both reflected an intertwining of languages and cultures. His father, Georg Iserin, had brought his wife, Thomasina de Porris, together with their young daughter, Magdalena, to Feldkirch from Lombardy shortly before Rheticus was born and baptized Georg Joachim. Most of his friends would call him Joachim, a common German name in the sixteenth century, though the Italian form of his first name, Giorgio, no doubt dropped more musically from his mother's tongue. When he visited Italy, that was the name by which he was known.

Feldkirch also stood at an educational crossroads. The town boasted a grammar school whose students had gone on to study at universities in Bologna, Paris, and Kraków.[1] Grammar schools were a unique Renaissance institution established to teach young boys—not only those destined for the priesthood—a knowledge of the Latin language as a foundation for their mastery of the liberal arts, starting with the trivium: grammar, rhetoric, and logic. Later, at a university, they could study the other four liberal arts, the quadrivium: arithmetic, geometry, music, and astronomy. The seven arts were supposed to be liberal in the root sense of "liberating." From his earliest days at the *Lateinschule* in Feldkirch,

CENTRAL EUROPE
in the 16th Century

Uppsala
Stockholm

SWEDEN

DENMARK

Hven
Copenhagen

North Sea

Baltic Sea

LITHUANIA

Königsberg

PRUSSIA

Danzig
Frauenburg
Olsztyn

Bremen

Elbe River

BRANDENBURG

Torun

SPANISH NETHERLANDS

Weser R.

Berlin
Wittenberg

Warsaw

POLAND

Rhine River

HOLY

SAXONY

Leipzig
Saalfeld

SILESIA

Oder River

Krakow

Vistula River

Frankfurt

Prague

ROMAN

Heidelberg

BOHEMIA

Nuremberg

HAPSBURG

CARPATHIAN MTNS.

Tübingen

Danube River

MORAVIA

ROYAL HUNGARY

LORRAINE

EMPIRE

BAVARIA

EMPIRE

Cassovia
(Kosice)

Basel

Munich

Vienna

Lindau

AUSTRIA

Zurich
Feldkirch

Budapest

SWITZERLAND

ALPS

HUNGARY

Milan

VENICE

SAVOY

MILAN

Po River

Padua
Venice

Danube River

Ferrara

APENNINES

Bologna

OTTOMAN

Florence

TUSCANY

Adriatic Sea

EMPIRE

PAPAL STATES

0 Miles 100 200

0 Kilometers 200

© 2006 Jeffrey L. Ward

Rome

Portion of Europe encompassing the life and travels of Rheticus.

Rheticus developed a Renaissance-humanist taste for the freedom that flowed from this radically integrated view of the branches of learning.

Also at the heart of the liberal arts, as one early educational theorist claimed, was the book—in Latin, *liber* [2]—and indeed it was for Rheticus, who was born amid the post-Gutenberg explosion of book production and book learning. A year after his birth, plans were laid for a civic library in Feldkirch. And his own father, Georg Iserin, who was town doctor of Feldkirch and himself a Renaissance man, possessed an impressive personal library as well as a collection of medical and alchemical equipment. Iserin took a direct hand in the education of his son, including, as Rheticus would attest, teaching him arithmetic—a subject not only highly practical but also promising "still greater advantages in all areas of life."[3]

As a child, Rheticus enjoyed considerable material advantages. His mother was of noble descent and a woman of means. His father found medicine to be a lucrative profession, no doubt augmenting his income, as was common for doctors in the sixteenth century, by selling astrological advice. Iserin, however, applied his skill, ambition, and opportunism to disastrous ends.

One standard account depicts Rheticus's father as a physician with something of a wild eye and a taint of foreignness who dabbled in fortune-telling and alchemical experiments—circumstances that allowed superstitious people to jump to conclusions about the source of his powers. The man with the books, the secret equipment, and the thirst for new knowledge began to look less like an ordinary citizen and more like a necromancing Dr. Faustus in league with the forces of darkness.[4] According to this story, a rumor circulated in Feldkirch that Iserin had a real live devil shut up in his laboratory, and he was brought up on charges of sorcery, an activity forbidden by the imperial penal code "upon pain of death."[5]

A better warranted account emerges from the actual transcript of Iserin's trial.[6] This document reveals that Rheticus's father faced a striking

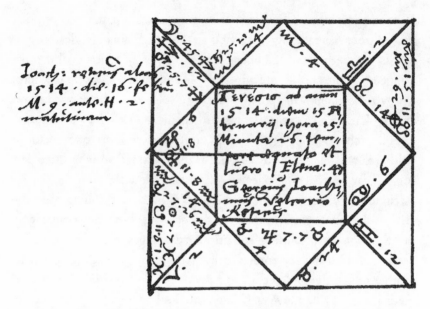

Horoscope of Rheticus drawn by his student Nicolaus Gugler.

number of charges, most of them alleging that he took dishonest advantage of his position as a doctor, even betraying his own patients. Evidence points to his having been a swindler and kleptomaniac who made off with money, gems, books, and silver chains and goblets from the houses (or pockets) of people with whom he was in a position of trust. Though it resulted in a guilty verdict, the lengthy trial seems in fact to have been impartial and to have followed due process, with the court making an honest effort to solicit mitigating evidence, what little there was of it.[7]

None of that made Iserin's sentence any less devastating; it left Thomasina herself frantic. According to a chronicle from the year 1530, she offered an enormous sum of money—"many thousand florins," perhaps ten times her son's eventual academic salary—in a desperate effort to win clemency for her husband. But in the end, the only mercy shown to Dr. Iserin was a concession regarding the manner of his execution.

Given his crime, he "should have been hanged."[8] But because he was a *Bürger* of Feldkirch and no mere common criminal, he was granted instead a privilege reserved for condemned citizens since the times of the Romans: He was publicly beheaded, with a sword.

Thomasina was thus violently bereft of her husband, and Magdalena and Georg Joachim of their father. They were at the same time stripped of their surname, for the criminal's condemnation also entailed a *damnatio memoriae*, a literal ignominy (= "no name"). The name Iserin was officially blotted out, and the remaining family reverted to Thomasina's maiden name, de Porris.[9]

Magdalena would eventually marry and take on her husband's name; Thomasina, when she remarried, would likewise acquire the surname of her new husband. But Georg Joachim, in keeping with a scholarly humanist custom of his day, eventually adopted a toponym, a name associated with one's birthplace. So proud was he of his home region—Rhaetia, an ancient Roman province encompassing corners of what are now Austria, Switzerland, and Germany—that he adopted a form of this Latin name as his own: Rheticus.*

Considering what happened to Georg Iserin, Georg Joachim Rheticus's enduring affection for his birthplace is striking. Despite his inner turmoil, there is no evidence that he protested or openly blamed anyone for his father's tragic end. He never spoke out against the court or its judgment. And given the strength of the evidence upon which Iserin's condemnation was based, the loss may have left Rheticus with an unfulfilled longing for a father whom he could properly honor and emulate.

The great wealth of Thomasina meant that soon after his father's execution in 1528 Rheticus was able to leave home and pursue the best

---

*In addition, Rheticus occasionally used the German version of de Porris, "von Lauchen." Both the Italian and the German surname mean "of the leeks."

education Europe had to offer. Already having begun his studies in Feldkirch—both at home and in the town grammar school—he traveled west to pursue them further at the cathedral school of the Frauenmünster Church in Zurich.

Rheticus threw himself into learning in this new setting on the western bank of the River Limmat where the waters of Lake Zurich drain northward toward the Rhine. Zurich was the center of the Swiss Reformation under Ulrich Zwingli, and Rheticus's main teacher here was Oswald Myconius (1488–1552), a onetime friend of Erasmus and notable humanist in his own right, as well as a reformer of high standing.*

From 1528 to 1531—from his fourteenth to his seventeenth year—Rheticus studied and dined with one of the leading intellectuals of his age in an atmosphere of both seriousness and excitement. Schools such as the Frauenmünster embodied a fusion of Renaissance and Reformation, a study of classical authors in both Latin and Greek complementing the inculcation of Christian prayer and practice. How deeply the latter emphasis influenced Rheticus is not clear, but significantly for him, Myconius taught mathematics in addition to the trivium, and Rheticus's commitment to that discipline deepened. In the wake of tragedy, Zurich was a place where Rheticus was nurtured, where he belonged, and where he made intellectual friends who would remain a kind of extended family as years went by. One of his closest friends was Conrad Gesner (1516–65), who remained in Zurich and became the most influential zoologist of his generation. The exceptional subsequent careers of both young men suggest that in Zurich, in addition to

---

*It was Myconius who had invited Zwingli to Zurich in the first place and who later became his first biographer. After he and Rheticus both left Zurich in 1531, Myconius went on to become chief pastor in Basel, offering haven to a number of foreign Protestants fleeing persecution, including William Farel and John Calvin from France and John Foxe from England, who there finished and published the Latin version of his *Book of Martyrs* (1559). Basel would also later serve as a major center for the dissemination of Copernicanism. The second edition of Rheticus's own *First Account* would be published there in 1541, as would the second edition of Copernicus's *Revolutions* in 1566.

The Rhinoceros as depicted by Rheticus's school friend the zoologist
Conrad Gesner (Zurich, 1551).

their immersion in language-learning and Reformed teaching, they were here inspired with a lasting love of science.

In the autumn of 1531 Rheticus left Zurich and returned to Feldkirch. Ironically, the man who now more than any other facilitated his higher education was the town doctor, the very position once held by his late father. This man, Achilles Pirmin Gasser (1505–77), became Rheticus's first major promoter.[10] An astronomer and historian as well as a physician, and only nine years older than his protégé, he had studied in Wittenberg, Vienna, Montpellier, and Avignon, and was prodigiously well connected. He would remain a faithful lifelong friend and a richly responsive scientific correspondent. But over the winter of 1531–32 he played the role of patron, nurturing Rheticus's interest in higher learning and, fatherlike, grooming him to leave home and enter the world beyond.

In 1532 eighteen-year-old Rheticus departed northward from Feld-kirch armed with Achilles Gasser's letter of introduction, bound for a place, in the words of one alumnus, "where the study of mathematics had always flourished": the University of Wittenberg.[11]

Founded in 1502 by Frederick the Wise, the University of Witten-berg had in fact been in existence only fifteen years when Martin Luther nailed his Ninety-five Theses to the door of the Castle Church in 1517, and thirty years when Rheticus enrolled there along with 290 other new students. The town itself had had roughly two thousand in-habitants at the founding of the university, a population that would al-most double by the mid-sixteenth century.[12]

Today the town officially bears the reformer's name: Lutherstadt Wittenberg. In 1532, on the strength of Luther's iron will and prodi-gious writings mass-distributed by the new technology of printing, the Reformation had already flooded out from this small town in Germany to the farthest reaches of Europe. In Zurich Rheticus had imbibed the excitement of a somewhat more recent current of Protestantism, but now he had arrived at its geographic and intellectual headwaters. Only two years after his arrival here, Wittenberg presses would produce the first copies of the full "Luther Bible"—an event that marked the debut of the German language as a respected medium of literary and learned communication.[13]

The very idea of Reformation was infectious, and Rheticus em-braced it. Lutheran fervor mixed with humanist scholarship—the translation and reinterpretation of ancient texts—not only produced monuments such as the Luther Bible but also nurtured a keen sense of discovery through reading. Given the long-standing analogy between the book of God's words (the Bible) and the book of God's works (the creation), there was also a natural analogy between the sets of tools used to interpret these two books: literacy and linguistic knowledge on the one hand, and mathematics applied to careful observations on the other. Not until 1623 would Galileo so clearly proclaim that "this

Sixteenth-century Wittenberg. The caption boldly begins: "Glorious city of God, seat and bulwark of true catholic [i.e., genuinely Christian] doctrine."

grand book, the universe, . . . is written in the language of mathematics."[14] But some of the roots of this idea go back to what Luther was doing in the 1520s and 1530s.*

The freedom to read not only the Bible but also Nature, and to reinterpret both for oneself, was supported by one of Luther's best-known doctrines: the "priesthood of the saints." "Saints" meant all believers, not just those whom the Church had canonized. This idea tended to undermine the authority of those who were officially priests—to cut out the middleman and offer ordinary Christians direct access to God and to God's Word. Indeed, so radical was this teaching that Luther himself never consistently encouraged its application, but it remained potent and infectious. If someone equipped with the tools of reading could reinterpret the text of either the Bible or the Book of Nature—independent of intervening layers of authority—whole new possibilities of understanding could emerge in the natural sciences as well as in theology. In this sense, what Rheticus and other mathematicians would

---

*It is not surprising, given his bold jurisdictional claim on behalf of mathematicians, that Galileo, himself a Catholic, would seem to his Vatican opponents to be embracing ideas tainted with Protestantism.

face in the long run was the challenge of codifying a language that let them read the book that Copernicus had pried open.

Two other practical changes that Luther brought about also helped enrich the story of Rheticus and Copernicus. These concerned music and marriage. In keeping with his blurring of the boundaries between priest and layperson, Luther encouraged clergy to marry and members of the congregation to sing. Luther loved music, proclaiming it—second only to the Word of God—worthy of the highest praise. Moreover, by promoting congregational singing, he turned passive observers into *participants* in this laudable dimension of creation. Pointing to "God's pure and perfect wisdom in his wonderful work of music," Luther declared: "Here [in polyphonic singing] it is indeed excellent how one and the same voice carries on singing the tenor, while at the same time manifold other voices wonderfully circle round it, moving playfully, exultantly, joyfully, and at the same time embellishing it, as if leading it on in a divine dance. And for those who in the least feel its effects, there is nothing more marvelous in the world. But those who are dead to its effects are boors indeed, fit to listen to nothing but shit-poets and the music of pigs."[15]

This vision of music would prove cosmically relevant to Copernicanism, for it offered an image of order and beauty within a system set in exuberant motion. In the new cosmology, one could celebrate the role of earth as a moving planet rather than a fixed point, for it meant that earth was participating in the music of the universe, in the "divine dance." Such too would be Galileo's response when he grasped anew our globe's planetary status after looking through his telescope for the first time: Earth was no longer "excluded from the dance of the stars."[16]

In addition, sung musical harmony offered a fitting metaphor for the cosmos itself, since it really is the same mathematics that underlie both polyphonic singing and the physical structure of the universe as a whole. Thus Johannes Kepler, the most Lutheran of Copernicus's early supporters, would assert in his great work *The Harmony of the World,* "I labor over the proportions and the whole harmonic panoply in music, in order to explain the reasons for the proportions of the celestial circles, and their eccentricities and their motions."[17] This thirst for harmony became part of Luther's and of Wittenberg's legacy to the world of Copernicus.

While Rheticus had little personal contact with the towering, superhumanly busy Luther, another major figure of the Reformation in Wittenberg had a powerful direct impact on his career. Philipp Melanchthon (1497–1560) became a humanist before he became a reformer. His great-uncle Johannes Reuchlin (1455–1522), who pioneered the study of ancient Greek in Germany, guided young Philipp's education in Pforzheim, near the Black Forest, and instilled in him a passion for classical languages. Melanchthon had enrolled at the University of Heidelberg at the age of twelve and completed his B.A. within two years, in 1511. The following year, at Tübingen, he continued his studies in Latin, Greek, philosophy, logic, mathematics, and astronomy, a subject that continued to fascinate him throughout his career. And in 1518, at the age of twenty-one, he was called to Wittenberg to serve as the second professor of Greek at any university in Germany.[18]

In Wittenberg Melanchthon became the pedagogical backbone of the university and Luther's chief lieutenant. Along with the priesthood of the saints, Luther promoted the marriage of priests, and this extended to professors. Often he would himself act as matchmaker, as he did in the case of the rather reluctant Melanchthon, whose first passion was.scholarship. Still, at the age of twenty-three he married Katharina Krapp, and so happy was their union that Melanchthon likewise became an avid proponent of marriage.

Melanchthon's educational influence, both beyond Wittenberg and beyond his own lifetime, is incalculable. His plans for the extension of humanist, liberal arts–based secondary schooling went into effect all across Germany, and his Latin and Greek school texts continued to be used there and in Scandinavia on into the eighteenth century. He is still widely known by the epithet *Praeceptor Germaniae*—"the teacher of Germany." In Wittenberg his lectures were attended by essentially the entire student body, numbering between four and five hundred. Finally, he wrote letters—of which an astounding 9,500 survive—to a network of correspondents across Europe, regularly promoting learning and learners. One of these protégés, starting in 1532, was Rheticus.

At his first meeting with Rheticus, Melanchthon sensed the young man's intellectual seriousness as well as his exceptional aptitude for mathematics. He would later declare that Rheticus was *natum ad Mathemata pervestiganda*—"born to study mathematics."[19] Rheticus's curriculum at Wittenberg also included astronomy, which was part of the mathematical discipline, as well as Greek, Melanchthon's chief specialty. Mathematicians such as Plato, Aristotle, Euclid, and Ptolemy were, after all, Greek authors. So Melanchthon saw to it that from every direction Rheticus absorbed humanistic learning along with instruction in things technical and mathematical. This effort further kindled Rheticus's passion for discovering the order, the causes, the beauty, and the interconnectedness of things.

Seeing his diligence and academic success, Melanchthon offered Rheticus the opportunity to remain in Wittenberg following receipt of his master of arts degree in 1536. His mathematics teacher, Johannes Volmar, died that same year, and the vacancy was filled by Erasmus Reinhold (1511–53). But Melanchthon soon arranged the establishment of a second professorship of mathematics at the university, and he named Rheticus to fill it.[20]

In the spring of 1536, two months after his twenty-second birthday, Rheticus took up his chair of mathematics and prepared to deliver, as was customary, an inaugural lecture. Such a performance would give the confident but still self-conscious young professor a formal public opportunity to justify his existence, to defend the value of his subject, and to display his eloquence.

Early in the lecture Rheticus recounted the ancient story of Athena, who, having been given a flute, an instrument recently invented by Hermes, plays it by the riverside. But suddenly Athena throws it away when, observing her reflection in the water, she sees her face "deformed" by her puffed-out cheeks—whereupon the discarded flute is taken up by the satyrs, who use it to make beautiful music.

The moral that Rheticus drew from this tale rests on the analogy between making music and doing mathematics. Both might be arduous, even perhaps disfiguring; but each is worth the effort it requires, and each nurtures the noble mind in search of truth: "It is characteristic of the honourable mind not to love anything more ardently than truth, and inspired by this desire, to seek a genuine science of universal nature, of religions, of the movements and effects of the heavens . . . If we are held by admiration and love for this perfect science, then we also need to value those elements of numbers and measures that provide access to the other parts of philosophy, although even on their own they are of great nobility and benefit."[21] Rheticus thus repeated the plea of mathematics teachers to their students over the centuries: to love the subject for its own sake as well as for its usefulness. Unfeigned love for this calling would continue to shape the career of Rheticus.

Rheticus's lecture also emphasized how vitally linked these two disciplines are: "There is no access to the science of celestial things except through arithmetic and geometry." This claim was poetically backed up by a citation from Plato, who in the *Phaedrus* "invents two kinds of souls, one of which he says is winged, the other has lost its wings. Then he says that those that have wings fly up to heaven and delight in meeting and conversing with God, and in the most beautiful sight of the heavenly ways . . . And, flying throughout all of heaven, these souls are charmed by the beauty of divine things and by the sweetness of knowledge and of the virtue of that admirable order."

Rheticus unfolded his own Christian interpretation of this parable: "The wings of the human mind are arithmetic and geometry." He exhorted his listeners, therefore, to "attach those wings." Rise up to the stars, those "true and enduring works of God, . . . carried by firm laws, because of some great purpose." Observe "the movements of the heaven" and so be reminded, in accordance with the Bible's astronomical passages, "that nature does not exist by chance, but is created and governed by an eternal mind."

Rheticus's lecture offered a rich blending of cultural materials— deeply rooted classical and humanistic affirmation of the liberal arts reinforced by love of poetic analogy—all interwoven with a Protestant sense of the individual's direct access to divine revelation, if this is supported by the right "texts" and "means of reading." Rheticus also betrayed a stern moral approach to learning: Schools do not exist "for the purpose of granting leisure to the idle for enjoying foolish pleasures." Infusing all these elements was Rheticus's own youthful exhilaration at new knowledge and fresh prospects, together with an unrelenting determination to pursue a unified, harmonious vision of the cosmos.

"Thrust away the idea of difficulty," Rheticus implored his students, "and approach this art with an open mind and some hope," so that you may be engaged by "the sweetness of things" and "feel the difficulty diminish by the very connection of the things." And remember: Without this art (mathematics), "there is no access to that most excellent part of philosophy dealing with heavenly matters."

Within two and a half years of taking up his professorship, however, Rheticus appeared to have had enough of Wittenberg. In October of 1538, right after the end of the summer semester, he asked Melanchthon, still serving as rector, for a leave of absence, and it was granted. Midway though that year, the atmosphere at the university had turned so sour that it was seriously impeding Rheticus's pursuit of "the sweetness of things."

The tension centered on one of Rheticus's former fellow students and drinking companions, Simon Lemnius (1511–50). Originally from the south of Rhaetia, Lemnius had come to Wittenberg in 1534, two years after Rheticus. Like Rheticus, he was promoted by Melanchthon, receiving his M.A. in 1535. But above all Lemnius was a skilled poet with a wicked eye for satire.* In Wittenberg Lemnius's main social circle was a collection of poetically inclined scholars that included not only Rheticus but also Rheticus's friend Caspar Brusch. Their poetry was broadly modeled on the erotic verses of Ovid—not at all in keeping with the norms expected by Luther, Melanchthon, or for that matter the secular authorities.

In the two years leading up to the summer of 1538, Luther had been in strenuous political and theological negotiations (aimed, unsuccessfully, at reconciliation) first with other Protestants and, then, in a highly preliminary way, with the Catholic Church itself. The tensions were taking their toll on Luther's relationship with his lieutenant Melanchthon, who to Luther looked increasingly a compromiser in the struggle with Rome. Occasionally he even appeared conciliatory toward Pope Paul III, to whom Copernicus some five years later would dedicate *The Revolutions*. Luther, for his part, had asserted that the pope was the Antichrist.

Lemnius's first offense was to publish—and to sell right outside the Wittenberg church door—a little volume containing two "books" of Latin poems, *Epigrams* (1538), vignettes in the classical style ranging from the satirical to the erotic.[22] Viewed through a humanistic lens, these might have been conventional enough to escape more than a

---

*Lemnius later went on to translate the entire *Odyssey* of Homer into Latin hexameters and to write his own epic, the *Raeteis*, celebrating the heroic deeds of his countrymen in the Swabian war of 1499—both of these enormous efforts appearing in the two years preceding his death of the plague at the age of thirty-nine.

small ripple of controversy. But Lemnius made one audacious mis-calculation. He dedicated his volume to Archbishop Albrecht of Mainz.

Albrecht was Luther's bitter enemy. In the second decade of the six-teenth century, it was he whose traveling indulgence-salesman, Johann Tetzel, had stirred Luther the young monk to take action against this perversion of the Gospel. After two decades of Reformation, Lemnius then had the gall to praise this same servant of the pope and publish right under Luther's nose a book in the archbishop's honor.

Luther and his closest allies, rather than objecting outright to the dedication of the *Epigrams*, suggested within a day of their publication on June 8 that the poems contained veiled slanders against certain hon-orable citizens of Wittenberg. Word of the growing uproar reached Lemnius by way of some of his closest friends, who the following day found him and advised him to flee. He hesitated, but in the middle of the night "a man of authoritative and erudite bearing"—many believe it was Melanchthon himself[23]—appeared before him and warned him again of his peril. And so, in the early morning hours of Monday, June 10, 1538, Lemnius fled for his life.

After Lemnius's departure from Wittenberg, a series of summonses went out calling for him to appear before the senate of the university. Prudently, he responded to none of them. Even while the summonses were going out, and only a week after the *Epigrams* were published, Luther took time in his Sunday sermon to read a prepared statement asserting that if Lemnius should be captured, "by all rights . . . he should lose his head."[24]

Although Lemnius erred in misreading the political mood in Witten-berg in 1538, Luther failed to foresee the consequences of an overreac-tion to this brash young writer. Melanchthon and other distinguished humanists looked on Lemnius's poems as relatively harmless and at worst deserving of academic reprimand.[25] More precisely, Lemnius had been harmless while still in Wittenberg. But in Germany beyond this stronghold of Protestantism there were many places where the demonstrated capacity for putting Luther into a towering rage was a prized commodity.

In September of 1538 Luther offered an estimation of Lemnius's skill and character in a Latin poem whose title indicates its tenor and contents: "A Dysentery upon Lemmy the Shit-Poet."[26] But Lemnius, exiled and embittered, now had, as far as Luther and Wittenberg were concerned, nothing to lose. In late September of 1538 he published another edition of *Epigrams* containing a new, third book, the tone of which is typified by an untitled poem in which he hurls the substance of Luther's doggerel back in the reformer's face: "Where formerly your crooked mouth spewed madness / It's now your arse from which you vent your spleen."[27]

Yet Lemnius's obsessive bitterness did not abate. Early in 1539 he published a prose treatise of self-defense (*Apologia*) in which he attempted to align himself with all that was moderate and reasonable in Wittenberg, in particular with Melanchthon and other creditable members of his circle. Among those who could purportedly vouch for his character he named Rheticus, "who is now there [in Wittenberg] a professor of astronomy."[28] Rheticus could scarcely escape guilt by association.*

Throughout his life Rheticus detested politics, intrigue, and anything else that threatened to distract him from his passion for higher learning—which in his case meant above all a love of beauty. The tension and vitriol that infused the university in the wake of the Lemnius

---

*Lemnius's other work of 1539 was also aimed at enraging Luther. Titled *Monochopornomachia* (*War of the Whoring Monks*), this burlesque portrayed Luther and his colleagues as lechers whose pathetic sexual performance could not satisfy the voracious appetites of their wives, who turned to younger men for their gratification. It lampooned Luther himself as a sexual deviant, and his beloved wife, Katie, as a woman having (with a play on her name) the night-wandering habits of a cat.

affair were merely ugly. Fortunately, however, by the time Lemnius's work appeared naming Rheticus as an associate, he was safely off on new scientific and intellectual pilgrimages elsewhere. For Rheticus, the winter of 1538–39 was a splendid time for an extended absence from Wittenberg.

CHAPTER 2

ROUNDABOUT ROAD
TO FRAUENBURG

In October of 1538, at the end of Wittenberg's summer semester, Rheticus left on a tour to visit key scholars in southern Germany. He carried with him a letter written by Melanchthon supporting this intellectual expedition. Both he and Melanchthon were happy to have him escape the tense circumstances in Wittenberg surrounding the Lemnius affair while at the same time deepening his knowledge of astrology, an "art" that interested both men and served as a driver of astronomical research throughout the sixteenth and seventeenth centuries.

On his journey south, Rheticus was accompanied by a young man named Nicolaus Gugler. It had long been common practice for a scholar to be assigned a famulus, a student who doubled as servant and academic assistant. It is a role remembered today through stories such as that of Dr. Faustus, with his underling Wagner, and of the sorcerer's apprentice. In 1536, at the age of fifteen, the precocious Gugler had attended Rheticus's astronomy lectures at the University of Wittenberg; he would also have been an ideal traveling companion because Rheticus's most important destination was Gugler's hometown, Nuremberg.

Whatever the initial aims of this mission, its greatest achievement would be to link Rheticus firsthand with the rich tradition of learning that Nuremberg embodied, a tradition whose main scientific substance was attached to the legacy of Johannes Regiomontanus (1436–76).

Viewed from the northern edge of Nuremberg, the slender spires of St. Sebald
adumbrate a pair of isosceles triangles against the sky.

Born in Franconia near the town of Königsberg ("King's Mountain"—
from which his own Latin name was derived), Regiomontanus gradu-
ated from the University of Leipzig and, at the age of sixteen, went to
Vienna to study with the famous astronomer Georg Peurbach
(1423–61). So close was their relationship that, in the words of an
early biographer, "the one was likened to a father, the other to a son."[1]
When Peurbach died at the early age of thirty-seven, he charged Regiomontanus from his deathbed to complete the half-finished work on
his *Epitome* of Ptolemy's *Almagest*.

The student then not only carried out this charge by his adopted "father" but also collected a further array of ancient manuscripts, intending to translate and publish them. He also aimed to test their theoretical claims against his own scientific observations. In pursuit of these goals he moved from Rome to Nuremberg in 1471, explaining his choice of location in these words: "I have chosen Nuremberg as my permanent home not only on account of the availability of instruments, particularly the astronomical instruments on which the entire science of the heavens is based, but also on account of the very great ease of all sorts of communication with learned men living everywhere, since this place is regarded as virtually the center of Europe."[2] Permanent ended up being only five years, for Regiomontanus died in 1476 at the age of forty. Part of his collection of manuscripts was dispersed or damaged, though many of his treasures passed into the hands of a series of Nuremberg editors, chief among whom was the cosmographer Johann Schöner (1477–1547). It was Schöner, sixty-one, whom Rheticus and Gugler first visited upon their arrival in Nuremberg in the autumn of 1538.

Schöner is best known today as an early maker of terrestrial and celestial globes, some of which still survive. From 1526 onward he was a professor of mathematics at the gymnasium (high school) in Nuremberg, which had been reorganized according to the new model of secondary education pioneered by Melanchthon. On the strength of Rheticus's connection with Melanchthon, Schöner honored the young scholars from Wittenberg with an invitation to stay at his own home. While they were there he could also have shown them a copy he had acquired of a world map produced in 1507 by Martin Waldseemüller. On its western edge appeared a continent identified—for the very first time—as "America."[3]

Schöner not only showed Rheticus hospitality but also encouraged him to see himself as a key player in that astronomical tradition—in that scientific community—reaching back to Peurbach and to the ancient world. More specifically, Schöner modeled for Rheticus the vital link between astronomy and trigonometry. Five years earlier, in August of 1533, Schöner had published one of Regiomontanus's greatest works, *On Triangles*, whose title page offered the work as a contribution "to the perfecting of astronomical knowledge," something impossible "without instruction in these things [i.e., triangles]." Or, as the book's preface declared, "No one can bypass the science of triangles and reach a satisfying knowledge of the stars."[4]

At the bottom of the title page of *On Triangles* is another clue to the importance of Nuremberg for the careers of Rheticus and Copernicus. The phrase below the geometric design reads, "All these things newly published, faithfully and with singular diligence, in Nuremberg at the house of Johannes Petreius." Petreius (1497–1550) was a skilled, ambitious Nuremberg printer-publisher who in the 1530s already had to his credit titles by (among many others) Erasmus, Luther, Pirckheimer, Melanchthon, Pico, Camerarius, and King Henry VIII of England.[5] By the time of Rheticus's visit, Petreius's partnership with Schöner as editor had made him the premier printer of scientific titles in all of Europe, and the unique confluence of these three men was about to make history.*

Rheticus's visit to Nuremberg in 1538 served as prelude to significant events. If Rheticus felt honored to enjoy the company of such eminent figures as Schöner and Petreius, they were no less impressed with his knowledge and character. In 1540, soon after seeing a copy of the *First Account* that had arrived for Schöner, Petreius would write to Rheticus reflecting on their time together in Nuremberg a year earlier. The letter recalled how Rheticus and Schöner had conferred "on the system of the motions which the wonderful celestial bodies display"; then, moving

---

*Rheticus may in part have undertaken his journey north to visit Copernicus as an "advance agent for Petreius." One of the gifts Rheticus gave Copernicus was a copy (still extant) of Schöner's edition of Regiomontanus's *On Triangles*. Rheticus thus in a quite literal sense was a bearer of the tradition of the science of traingles. See N. M Swerdlow, "Annals of Scientific Publishing: Johannes Petreius's Letter to Rheticus," *Isis* 83.2 (June 1992): 270–274 (p. 271).

DOCTISSIMI VIRI ET MATHE-
maticarum diſciplinarum eximñ profeſſoris
IOANNIS DE RE-
GIO MONTE DE TRIANGVLIS OMNI=
MODIS LIBRI QVINQVE:
Quibus explicantur res neceſſariæ cognitu, uolentibus ad
ſcientiarum Aſtronomicarum perfectionem deueni-
re:quæ cum nuſquã alibi hoc tempore expoſitæ
habeantur,fruſtra ſine harum inſtructione
ad illam quiſquam aſpirarit.

Acceſſerunt huc in calce pleraꝗ D.Nicolai Cuſani de Qua
dratura circuli, Deꝗ recti ac curui commenſuratione:
itemꝗ Io.de monte Regio eadem de re ἀλεγκ-
κά, hactenus à nemine publicata.

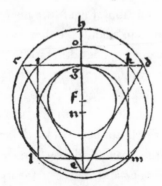

Omnia recens in lucem edita, fide & diligentia
ſingulari, Norimbergæ in ædibus Io. Petrei,
ANNO CHRISTI
M. D. XXXIII.

Title page of Regiomontanus's On Triangles *(1533).*

beyond pleasantries, Petreius commented on Rheticus's journey "to the farthest corner of Europe" to visit Copernicus: "a distinguished gentleman whose system, by which he observed the motions of the heavenly bodies, you related to us in a splendid description." Although Copernicus "does not follow the common system by which these arts are taught in the schools," Petreius continued, "I consider it a glorious treasure if some day through your urging his observations will be imparted to us, as we hope will come to pass."[6]

Petreius's flattery of Rheticus was partly politeness, partly an effort to drum up business for his press. Nonetheless, his praise authentically captures a passion for beauty that characterized Rheticus's entire career: "My Joachim, I congratulate your discernment because, although by the example of others you could very easily pursue the lucrative arts, you have set yourself a different course in order that you may provide yourself with true and lasting knowledge of the most beautiful arts."[7]

The warmth of Petreius's commendation was reciprocated by Rheticus's own ongoing cultivation of his Nuremberg contacts, Schöner in particular. During Rheticus's adult life there was, apart from Copernicus, only one other man whom he addressed in print as father, and that was Schöner. Rheticus would begin the *First Account,* in keeping with its epistolary form, with a greeting to Schöner, "as to his own revered parent." And he would close it with the promise always to honor Schöner "like a father" whose "paternal love" (*paternus amor*) had "impelled [him] to enter this heaven" of new knowledge.

In addition to Schöner and Petreius, Georg Hartmann and Andreas Osiander, both Nurembergers as well, would prove influential in the career of Rheticus. Osiander (1498–1552) was a theologian and keen amateur mathematician. During the previous decade he had played an important role in bringing the Reformation to Nuremberg as well as in encouraging young Duke Albrecht of Prussia to convert to Protestantism.

The mathematician Georg Hartmann (1489–1564), whom Rheticus

would honor as a scholar "famous for both his learning and his character,"[8] was a maker of astronomical instruments, some of which he gave or sold to important figures such as Melanchthon and Duke Albrecht of Prussia. As discoverer of the inclination of the magnetic needle, he was also a major contributor to the science of terrestrial magnetism.[9] Rheticus particularly valued Hartmann's interest in the connection between trigonometry and astronomy. When Rheticus published *On the Sides and Angles of Triangles*, he would dedicate it to Hartmann, declaring, "The science of triangles, as you know, is useful in all things pertaining to geometry, but especially in astronomy."[10]

In the waning days of 1538 Rheticus and Gugler continued farther south to Ingolstadt, where they met Peter Apian (or Apianus, 1495–1552). Like Schöner, Apian was a maker and collector of maps. In 1520 he had produced a world map based on the now-famous 1507 original by Waldseemüller. In 1524 he published a major work titled *Cosmographia*, ambitiously subtitled *Description of the Whole Globe*. This work was revised in 1529 by Gemma Frisius, the teacher of Gerardus Mercator and later (in 1541) one of the earliest avid readers of Rheticus's own *First Account*.

Like Rheticus and Gemma Frisius, Apian was keenly interested in the practical application of triangles. In 1534 he had published the first sine tables calculated for every minute of arc, which he offered for use in navigation, architecture, and astronomy. The publication that made him a rich man was his 1540 *Astronomicum Caesareum*, dedicated to the emperor, Charles V. This work recorded for the first time that the tail of a comet always points away from the sun. For his part, Charles gratefully rewarded Apian with the privileges of imperial mathematician, including the right to declare illegitimate children legitimate and to award higher degrees.[11]

Leaving Ingolstadt, Rheticus and his young famulus traveled farther south and west to the university town of Tübingen. Here they were received by one of Germany's foremost humanists and a close friend of Melanchthon, Joachim Camerarius (1500–1574). A number of works translated or edited by Camerarius had been printed by Petreius in Nuremberg. Camerarius had become headmaster of the gymnasium in Nuremberg at the age of twenty-six and then in 1535 professor of Greek at Tübingen. Rheticus and Camerarius would maintain a warm personal correspondence on into the 1560s.*

Another lifelong friendship that began in these early months of 1539 was that of Rheticus with Camerarius's son, Joachim Jr. (1534–98). Only fourteen at the time of Rheticus's visit, the younger Camerarius went on to study in Wittenberg under Caspar Peucer (1525–1602), a subsequent student of Rheticus and son-in-law of Melanchthon. Joachim Jr. became an influential herbalist, founded a school of medicine in Nuremberg in 1564, and in 1581 inherited the botanical manuscripts of Rheticus's old Zurich classmate Conrad Gesner. Rheticus's last extant letter, dated Kraków 1569, was written to Camerarius Jr. in Nuremberg and described with great frankness the painful bladder condition of a mutual acquaintance.[12]

In early 1539 Rheticus continued farther south to Feldkirch. It was not quite seven years since Rheticus the relatively raw youth had left home to begin his university education at Wittenberg. Though still young and only turning twenty-five, he was already a professor, a man with whom some of the great minds of Germany had pondered the heaviest scientific issues of their day. Yet Rheticus did not forget where he came from, in spite of Feldkirch's association with his father's judicial murder; nor could he neglect the chance to renew his friendship with one of his most faithful living patrons, Achilles Gasser.

---

*Two years after Rheticus' visit in Tübingen, Camerarius would move again, to the University of Leipzig, where he would be instrumental in securing a position for Rheticus himself in 1542. When Petrus Ramus wrote from Paris in 1563 (see pp. 174–175), he would suggest that Rheticus send his reply via "the most learned Mr. Camerarius, your dearest friend."

When he visited Gasser in the late winter or early spring of 1539, Rheticus exhibited what came to be one of his most endearing habits. He arrived bearing books—books that bespoke scientific community as well as affection and respect. In this case, one of his literary presents was a book on astrology written by Schöner, his new mentor in Nuremberg, and printed there by Petreius. Gasser in turn showed his young guest the manuscript of a thirteenth-century work by Petrus Peregrinus titled *Epistola de magnete,* which Gasser would subsequently publish and Rheticus would cite in his *Chorography* of 1541.

Rheticus's conversations with great minds over the winter of 1538–39 so sharpened his thirst for truth and beauty that he resolved to travel to the far north to visit an aged amateur astronomer named Copernicus and to hear for himself how seriously he maintained, as was rumored, that the earth was in motion and two thousand years of natural philosophy was crucially in error.

So when Rheticus returned to Wittenberg early in May of 1539 after almost seven months away, he scarcely broke his northward momentum. In little more than a week he prepared for his further pilgrimage. He gained permission from the academic head of the university—Caspar Cruciger, a longtime supporter who had recently succeeded Melanchthon as rector—for an extension of his leave of absence. At the same time he arranged for a replacement for Nicolaus Gugler, who had stayed behind in Tübingen. His new famulus was Heinrich Zell, who had arrived in Wittenberg in 1538 just before Rheticus's departure for Nuremberg. Zell was three years older than Gugler and only four years younger than Rheticus himself. More than a mere academic servant, Zell would prove to be a friend and exceptionally competent companion, and the journey northward would also be the making of Zell's own significant scientific career.

Rheticus and Zell set off together from Wittenberg, the very heart of the Protestant Reformation, bound for Frauenburg, the cathedral town of the Catholic episcopal province of Varmia. Their journey of many days first led eastward, across the River Oder to Posen (Poznań). From here, on May 14, Rheticus sent a letter to Schöner in Nuremberg informing him that the Copernican quest was successfully under way.[13]

Continuing their journey, the travelers also stopped in Toruń, Copernicus's birthplace on the northern, right bank of the Vistula, a cosmopolitan trading center and thriving member of the Hanseatic League. From there they either took transport northward, downriver, partway toward the Baltic, and then continued eastward again by land, or else traveled cross-country from Toruń, bearing east by northeast. Either route would have brought them eventually to Elbing, the last major town before their final destination of Frauenburg.

Frauenburg (Frombork) as depicted in a seventeenth-century copper etching.

Almost a day's journey east of Elbing, Rheticus and Zell branched off onto a narrower road leading north. Before long, their way emerged abruptly from the wood, and the travelers found themselves suddenly there, at the edge of a small, raggedly triangular town lying at a slight northward tilt along the southern fringe of Vistula Bay. Through the middle of the town flowed a diminutive river that, just before it entered the bay, widened to form a compact harbor where Frauenburg fishermen moored their open flat-bottomed boats and cleaned their eels.

Ninety degrees to the west and a quarter of a mile from where the road had deposited the travelers upon the coast, a hill rose up behind the town, seeming almost to lean over it. Imposingly atop this hill stood a massive redbrick cathedral and the fortification (*Burg*) that gave the town part of its name. Copernicus, in his somewhat ironic way, occasionally referred to this place using a Greek equivalent of Frauenburg: *Gynopolis*. He would also call it, with a touch of melancholy, "this remotest corner of the earth."

But in that late May afternoon of 1539, Rheticus, as he gazed for the first time at the Frauenburg cathedral backlit by the slowly westering sun, was looking in toward the center of a new universe.

CHAPTER 3

VITA COPERNICI

In his youth and early middle age Nicolaus Copernicus had enjoyed a rich range of correspondents and acquaintances. But given the abstruse cast of his intellect, by May of 1539 not many were left with whom he could truly converse. Three and a half continuous decades in a far-off province like Varmia had accustomed him to a certain lack of connectedness with the wider world.

There were still many whom he loved and respected—not only fellow canons (high-level ecclesiastical and provincial administrators, know collectively as the chapter), but also real friends like Tiedemann Giese and Georg Donner, as well as Copernicus's sometime housekeeper Anna Schilling. There was also Duke Albrecht in Königsberg, whose acquaintance with Copernicus seemed to grow deeper every year. Yet a certain routine, even boredom, was apparently part of what it meant to lead a peaceful existence. Varmia had not experienced any armed conflict for a good number of years, and Copernicus had seen more than enough of that to be thankful for its absence.

Unfortunately, though, this time of peace seemed to offer the sitting bishop of Varmia, John Dantiscus, ample leisure for persecuting his friends, including Copernicus, whom he still claimed to "cherish . . . as a brother."[1] Dantiscus had entered the priesthood at the late age of forty-five, after leaving a long diplomatic career that had taken him to

Austria, the Low Countries, and Spain, where he had also left an illegitimate daughter. In 1530 he had become bishop of Kulm, a position which in 1537 he relinquished to Copernicus's friend Giese so as to be able to occupy—and prevent Giese from occupying—the richer bishopric of Varmia.

Then, apparently hoping somehow to atone for his own past sins, from his seat in Heilsberg he mounted a campaign against the Frauenburg canons' employment of female housekeepers. With regard to this issue Copernicus had long exercised a polite passive resistance, saying that he would obediently break off contact with Anna Schilling while in fact—as Dantiscus alleged—continuing to have "secret trysts" with "his harlot."[2] In March of 1539 Dantiscus also published an edict against heresy in which he stipulated a one-month deadline for the permanent expulsion from Varmia of anyone not adhering to the prescribed Catholic Church order. And he would soon pass on a pointed warning to Copernicus not to be led astray by those under suspicion of the main heresy that he had in mind: Lutheranism.[3]

In late May of 1539 Copernicus was informed that a young mathematics professor had arrived at Frauenburg and wished to speak with him. He agreed to a meeting, and in their first encounter Copernicus discerned, if he had not been informed already, that the man's name was Georg Joachim Rheticus, that he had come all the way from the Lutheran hotbed of Wittenberg, and that the purpose of his journey was fortunately astronomical, not theological. During the previous year Rheticus had heard that Copernicus was framing a new conception of the heavens, one that asserted the motion of the earth, and he now respectfully requested that Copernicus introduce him to this novel theory.

It was not the first time Copernicus had been asked to divulge his unique cosmology. Fifteen years earlier, in 1524, at the urging of a former Kraków acquaintance named Bernard Wapowski, secretary to the king of Poland, Copernicus had offered his (negative) evaluation of a book on astronomy by the Nuremberg mathematician Johann Werner. In that letter, known simply as "Against Werner," Copernicus had promised Wapowski that at some later point he would set forth his views "concerning the motion of sphere of the fixed stars."[4] Much later, in 1535, Wapowski wrote to a Viennese diplomat informing him that Copernicus was preparing a new account of the movements of the planets, according to which "some motion must be granted to the earth."[5]

Copernicus's earth-moving notions had also reached the highest circles in Rome. A young German named Johann Albrecht Widmanstetter, a papal secretary, spent an afternoon in 1533 in the Vatican gardens delighting Clement VII with his explanation (as he himself put it) of "the Copernican opinion concerning the motion of the earth."[6] When Pope Clement died two years later, Widmanstetter became secretary to Cardinal Nicolaus Schönberg and likewise informed him of Copernicus's ideas. Schönberg in turn wrote directly to Copernicus in 1536 pleading for a fuller account of his "new cosmology," of which he already knew the outlines.

> In it you teach that the earth moves; that the sun occupies the lowest, and thus the central, place in the universe; that the eighth heaven remains perpetually motionless and fixed; and that, together with the elements included in its sphere, the moon, situated between the heavens of Mars and Venus, revolves around the sun in the period of a year. I have also learned that you have written an exposition of this whole system of astronomy, and have computed the planetary motions and set them down in tables, to the greatest admiration of all. Therefore with the utmost earnestness I entreat you, most learned sir, unless I inconvenience you, to communicate this discovery of yours to scholars, and at the earliest possible moment to send me your writings on the sphere of the universe together with the tables and whatever else you have that is relevant to this subject.

Copernicus took this letter seriously enough that he saved it and eventually had it printed as a headnote to *The Revolutions* in 1543, proving that important people had urged him to publish his work. (See the illustration on p. 111.) In 1536, however, whether because he was too busy or because he knew he could not yet scientifically prove his theory, Copernicus had not responded to Schönberg's request. Then in 1537 the cardinal died, as had Wapowski in 1535, and the voices of the curious seemed to fall silent.

Yet here in this late May of 1539, Rheticus had arrived as the very embodiment of intelligent curiosity. In contrast to Schönberg and Wapowski, not only was Rheticus young and vigorous; he was also Protestant, though this fact clearly did not bother Copernicus at all. What is somewhat more surprising is that it also apparently did not bother Bishop Dantiscus, who had just two months earlier issued his edict against Lutherans.

The logic of the bishop's actions, however, was most likely rooted in animosity rather than religiosity. The man he was especially keen on prosecuting for Lutheranism was one of Copernicus's much admired fellow canons in Frauenburg, Alexander Scultetus, whose real crime was to have opposed the election of Dantiscus first as canon and then as bishop. Rheticus, despite his Lutheranism, had no such "past" as far as the vindictive bishop was concerned. In addition, Rheticus likely carried with him letters of commendation from prestigious persons such as Melanchthon in Wittenberg, Schöner in Nuremberg, and possibly Peter Apian in Ingolstadt, a Catholic who was soon to become Charles V's imperial mathematician.

As for Copernicus himself, what moved him most deeply and evoked from him a memorably warm welcome was not only Rheticus's curiosity but also his devotion—devotion to scientific understanding of the universe, devotion that had carried him all the way from Nuremberg and Wittenberg to these northern shores, and now, almost as if they had known each other for a very long time, devotion to *him*, Copernicus, as a man as well as a scientist. Copernicus in turn, as he had never in his life done before, opened his doors, his mind, and his life to a young man—his first and only student—who ever afterward would refer to Copernicus affectionately and simply as "my teacher."

Rheticus spent most of the next two years and four months, until September of 1541, with Copernicus. During this time Rheticus composed a biography, a *Vita Copernici*. Writing to Rheticus in 1543, after the publication of *The Revolutions* and the subsequent death of Copernicus, Tiedemann Giese would mention the biography—"your elegant life of the author, which I once read." Tragically, this work is lost, and there is no trace of anyone other than Giese having read it. The fact that Rheticus did write it, however, suggests that he carefully observed and absorbed the life as well as the work of his new teacher. His biography would have included not only things related directly to science but also elements that resonated or even noticeably contrasted with his own experience.

Throughout his life Rheticus was actively concerned with cultivating "fathers" and with managing the benefits and responsibilities of patronage. He would have been fascinated with Copernicus's main father figure, a formidable character by the name of Lucas Watzenrode (1447–1512). The brother of Copernicus's mother, Barbara Watzenrode, Lucas was a man most of Copernicus's biographers like to dislike—though it is not always clear why. Some report that Uncle Lucas was dour, and that no one had ever seen him laugh.[7] Certainly, he was a man who usually got what he wanted, and who embodied literally the practice of nepotism, "favoritism toward nephews."

Just twenty-five when Nicolaus Copernicus was born on February 19, 1473, Lucas was already something of a Renaissance man, having studied at the great universities of Kraków, Cologne, and Bologna, where he received a doctorate that same year. Shortly thereafter he returned home to his birthplace of Toruń, a lively, cosmopolitan, mercantile town on the Vistula River, where for a time he taught school.

During his first year or so back in Toruń, Lucas seems to have been somewhat at loose ends. He soon took up with "a pious virgin" (in the words of a rather equivocal early chronicler)[8] who happened to be the daughter of his school's principal. The result of this liaison was an illegitimate son named Philipp Teschner, who decades later, after Lucas became bishop and ruler of Varmia, served as mayor of Braunsberg, a Varmian town not far to the east of Frauenburg. Lucas always looked after his own. Not surprisingly, his liaison with the principal's daughter also resulted in a sudden career change, away from school teaching— and toward the church. Starting in 1475, Lucas began accumulating a series of appointments as canon, the most notable one being in Frauenburg in 1479. This was followed in 1489 with his election as bishop of Varmia, a position he held until his death twenty-three years later, on March 29, 1512, just after Copernicus turned thirty-nine.

But back in 1483, upon the death of Nicolaus Copernicus Sr.— Copernicus's own father—Uncle Lucas had stepped in to care for and educate the children of his brother-in-law: two girls and two boys, Nicolaus Jr. being youngest of the four. Copernicus turned sixteen on the day his uncle was elected bishop of Varmia, and Lucas planned that his nephews—both Nicolaus and his older brother, Andreas—should follow in his footsteps as canons in his own cathedral chapter in Frauenburg. The first requirement, though, was that they attend a university.

The initial stage of this plan involved the Copernicus brothers' enrolling at the great Jagiellonian University in Kraków, Lucas's own alma mater. By winter 1491, Nicolaus and Andreas had journeyed south to begin their *studium* in the Polish capital, where their names can still be read in the matriculation records for that year ("Nicolaus Nicolai" and "Andreas Nicolai," in that order). Nicolaus left Kraków less than four years later without receiving a degree, traveled in Germany and nearer home in Prussia for a time, returning to Frauenburg briefly, but in the autumn of 1496 he began studies in canon law across the Alps in Italy, at the University of Bologna, where his Uncle Lucas had received his doctorate.

Copernicus returned to Frauenburg in the summer of 1501 and applied for support for a further period of study in Italy. His chapter granted this request on condition that he pursue studies of long-term practical benefit to his community, so Copernicus agreed to study medicine. He did this for about two years, at Padua, the same university that Rheticus's father most likely attended.[9] However, since Copernicus could not earn an M.D. degree in two years, he elected to round out his Italian sojourn by taking a doctorate in canon law, on May 31, 1503, at Ferrara.*

For nearly a dozen years Copernicus had enjoyed a high-quality international formal education with both the direct and indirect support of his uncle Lucas. And in 1503 it was time for another kind of education. Instead of taking up his canonry in situ in Frauenburg, Copernicus joined his uncle in the episcopal palace in Heilsberg and assisted him in his strenuous political maneuverings on behalf of his province, the diocese of Varmia.

As seen from a map of Prussia in Copernicus's day, Varmia was a keystone province entirely surrounded by East Prussia. Looked at on a larger canvas, it lay roughly between Royal Prussia, which like Varmia owed fealty to the king of Poland, and East Prussia, which was ruled by the Teutonic Knights. This latter order, dating from the crusades, had colonized Prussia over a period of three centuries and for much of Copernicus's life continued to act aggressively in relations with its neighbors.

Much of Lucas Watzenrode's time was taken up with efforts to preserve the relative self-government of Varmia (*self-government* meaning

---

*Canon law is the system of law governing the Roman Catholic Church and its institutions. Varmia, because it was an ecclesiastical province, was under canon law, just as most Western jurisdictions today are under either civil law or common law.

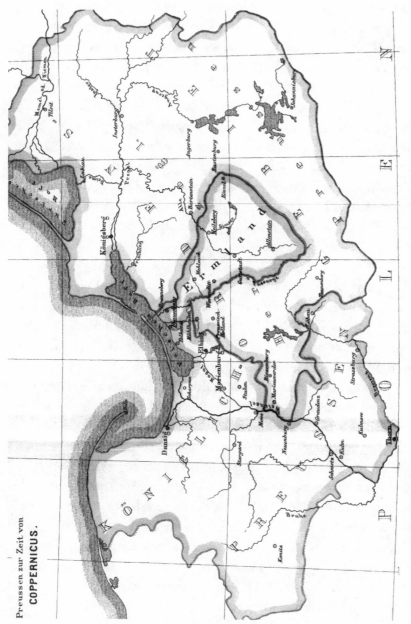

Northern Poland and Prussia, with the province of Varmia identified by its German name, Ermland.

effectively government by the bishop himself). This meant holding the Teutonic Knights at bay. Understandably, Watzenrode was no favorite of the Knights. According to their own chronicles, and in keeping with their rather peculiar style of piety, they offered up daily prayers that God might remove "this devil incarnate" from the world.[10]

Under the terms of the Treaty of Kraków in 1525, however, the whole political situation in Prussia changed. The Order of the Teutonic Knights was dissolved and a secular dukedom established owing loyalty to the Polish crown. The former grand master of the Knights, Albrecht of Brandenburg, became duke and promptly converted to Protestantism. Having widened the scope of his interests to include theology and the liberal arts, the duke ironically was now at peace with his Catholic former enemies.

There are still other reasons why Copernicus's young visitor would have been fascinated by the character of Uncle Lucas. Watzenrode, like Rheticus's own father, embodied some of the notable ideals of Renaissance humanism—to which both Rheticus and Copernicus aspired. When Copernicus decided not to follow in his uncle's footsteps in his pursuit of a career, he made efforts to blunt Lucas's disappointment by demonstrating that he still remained true to the humanistic culture in which Lucas had raised him.

In the sixteenth century it was common practice for an author to use publication as a tool of diplomacy, to attempt to ingratiate himself to a patron (or hoped-for patron). In this way Copernicus dedicated *The Revolutions* to the pope, and Rheticus likewise dedicated a number of his works to men of status, wealth, or power. But in Kraków in 1509 Copernicus published, in humanistic fashion, a Latin translation of a literary cycle of letters by a late sixth-century Byzantine writer named Theophylactus Simocatta. This was the first translation of a Greek author ever published in Poland, and Copernicus dedicated it to his uncle Lucas.

Copernicus's letter of dedication to his patron was in most respects purely conventional, though it offered some understated personal touches. Copernicus observed, for example, that there are different types of people—not only those who prefer "levity and bright fantasy" but also those inclined to "seriousness and grave realism"[11]—comments that may have acknowledged his uncle's stern character while putting it in a kindly perspective.

Among the letters themselves, one stands out in light of how Copernicus's scientific career eventually unfolded. Letter 31, "From Hephaestio to Thales," borrows astronomical language to describe the show put on by the peacock: "When it opens out its round shapes, it resembles the heavenly display, with the eyes on its feathers duplicating the form of the stars." But this open display is offered as a contrast to the career of the reluctant Thales, still today generally credited as the very first scientist, yet who left no writings: "You [Thales], on the other hand, sit on your writings, hide your accomplishments, and setting you works aside in obscurity, you disregard us who are deprived of so great a boon."[12]

When it later came to *The Revolutions,* Copernicus himself would in fact face the same highly practical issue of whether merely to sit on his writings. Fortunately, Rheticus appeared on the scene to play the role of an enthusiastic Hephaestio, and Copernicus would see that it was better to escape the obscurity, and the rebuke, of Thales.

The Theophylactus translation was also symbolically significant simply because it was a translation.[13] It stands as a reminder of the unique age in which Copernicus lived and of the profound educational values he had absorbed, values that would unite Copernicus and Rheticus despite their many differences of age, culture, and character.

Lucas Watzenrode, Copernicus, and Rheticus were "Renaissance men" not merely because they lived at a certain period of history but because they were immersed in ways of thinking that typified Renaissance humanism. Twenty years before Copernicus was born, Byzantium fell to the Turks, and consequently whole libraries embodying the literary and philosophical treasury of the ancient world were displaced westward into Italy. The last third of the fifteenth century then witnessed the flourishing of a textual industry that rendered into Latin many ancient writers whose works were available for the first time since the fall of Rome a thousand years earlier. The overarching characteristic of Renaissance humanism was the value it placed on ancient learning as a fund of new knowledge—or, more correctly, of old knowledge that could be newly discovered.

This enthusiasm specifically shaped the career of Rheticus, and on a larger scale it formed the main connection between the Renaissance and the Reformation. Both of these movements were based on the recovery and reinterpretation of ancient texts. According to the cliché, "Erasmus laid the egg that Luther hatched." Erasmus, seven years older than Copernicus, edited the Greek New Testament that Luther, ten years younger than Copernicus, translated into German (1522). Furthermore, with Luther's Bible or newly edited classical works, translation meant more than simply a mechanical rendering of texts into a familiar language. It invited wholesale reinterpretation and conveyed an exhilarating sense of access to newly rediscovered sources of truth.

Renaissance translation was actually a closer cousin to modern science than it may appear at first glance. Both translation and science search for underlying consistency in the phenomena they study. Astronomers attempted to "save the appearances"—to interpret them in a self-consistent manner, to explain a multiplicity of apparently haphazard phenomena, connecting them to an underlying harmony. Similarly, the translator struggles with the "surface" of the text and attempts to present the often confusing expression of ideas in a foreign language in a clear, coherent manner.

As St. Bonaventure wrote in the thirteenth century, "The whole world is a shadow, a way, a trace; a book with writing front and back."[14]

It was the task of the natural philosopher to translate and interpret its language in a consistent and illuminating manner, just as the translator of Plato or the Bible or Theophylactus aimed to make sense of the original and to communicate it to a wider audience. In *The Revolutions* one of Copernicus's deepest motivations for developing his sun-centered model was his belief that earlier interpreters of nature had produced a "translation" that was incoherent and aesthetically unappealing—one that did not do justice to the skill of the original Author-Creator.

At an individual level, translation also forces the translator to examine his or her own usual ways of thinking. Learning a language often involves moments of "defamiliarization"; the student discovers there is no word for something thought to be familiar to everybody, or discovers there is a name for something that he or she can hardly imagine. The act of translation thus exposes the strangeness of another "world," and perhaps the contingency or arbitrariness of one's own. Copernicus the astronomer can similarly be viewed as a retranslator of the text of the world itself, though at first his translation seemed, even to him, strange and absurd. Rheticus arrived, however, and soon mastered the new idiom. And suddenly Copernicanism appeared, to its founder as well, as a living language worth passing on to posterity.

Still other dimensions of Copernicus's career would have caught Rheticus's attention and afforded him useful insights into the workings of the astronomer's mind. One of these involved Copernicus's work in economics—his contribution to currency reform. His *Essay on the Coinage of Money* offered the first formulation of the principle that "bad money drives out good," now known as Gresham's law, though Thomas Gresham (1519–79) was not actually its first discoverer.[15]

Today monetary exchange is something in which essentially nothing of concrete value changes hands. But in sixteenth-century Prussia there was no paper money, and commerce was carried out by means of coins made from a mixture of copper and silver. Unlike paper money, the medium of exchange itself contained value.

Copernicus observed that, over time, the proportion of silver to copper in coins had been reduced. The effect was that different coins of the same face value had different actual values. If a person had two coins and knew that one of them contained a higher proportion of silver than the other, he or she would naturally use the coin containing less silver for making purchases. Thus "good" money would tend to be taken out of circulation because of the presence of "bad" money. Copernicus recognized that this phenomenon occurs whether a government introduces a new, intrinsically more valuable currency without withdrawing the old, or introduces a new, less valuable one: "It is not in the least advisable to introduce a new, good coinage while an old, debased coinage remains in circulation. How much worse was this mistake, while an old, better coinage remained in circulation, to introduce a new, debased coinage, which not only spoiled the old coinage but, so to say, swept it away!"[16]

Yet the problem was not merely with money itself in isolation. If coinage were of variable value, a merchant would also be suspicious when accepting a coin in exchange for the product he was selling, and indeed a person might be reluctant to sell at all without being certain of the intrinsic value—the silver content—of the coins being offered. Copernicus could see that the debasement of the Prussian currency was affecting both commerce within the country and trade beyond it.

Early in his discussion of money, Copernicus asserted that "coinage is . . . a measure of values. A measure, however, must always preserve a fixed and constant standard." It must be of "an invariable magnitude."[17] In economics Copernicus insisted on constancy, order, and invariability—just as in cosmology he would extol the "orderly arrangement" and "marvelous symmetry" of the universe.[18] In practical terms, he accordingly sought a unified and stable currency by proposing that coins of constant value be minted at only one location, or in any case at no more than two locations. For "multiplicity interferes with uniformity."[19]

In theoretical terms, Copernicus was intent on finding or establishing a "common measure" (a phrase Rheticus would introduce into the literature of astronomy)[20] whereby to judge and explain apparent change. According to Copernicus, if a currency fails to be a reliable measure, "gold, silver, food, household wages, workmen's labor, and whatever is customary in human consumption soar in price. But, being inattentive, we do not realize that the dearness of everything is produced by the debasement of the coinage . . . [Formerly] grain and produce were bought in Prussia with a smaller number of coins while sound money was still being used. Now, however, as it is being debased, we experience a rise in the price of everything."[21]

What Copernicus recognized, in economics as in astronomy, was a kind of relativity. Just as he and Rheticus would later show how the apparent movement of the sun can be explained in terms of the actual movement of the earth ("granting immobility to the sun, we exchange earthly movement for solar movement," etc.),[22] so he saw that in economics the apparent rise in the cost of goods actually marked a depreciation in the value of the currency. In both fields it was necessary to be "attentive" and realize that things may be other than—even the reverse of—what they seem.

When Rheticus set out to visit Copernicus in 1539, his motivation was exclusively mathematical and astronomical. After his arrival in Frauenburg, however, he soon discovered that his new teacher and father figure—like his own actual father—was a medical doctor. And by the standards of his day, Copernicus was reputed to have been an exceptionally good one.

Details of Copernicus's medical education and practice offered Rheticus some clues regarding how his teacher's mind had developed. All of the learned disciplines at that time were more intertwined than they are now—and this was perhaps especially true for medicine and astronomy, which often blurred together with astrology. One idea in particular informed this relationship. About 400 BC the Greek philosopher Democritus declared that "man is a small 'ordered world' [*kosmos*]."[23]

From then on, the idea of a fundamental correspondence between the universe as a whole and the individual person, between the *macrocosm* and the *microcosm*, exerted a tremendous influence in many areas of human thought. It seemed only logical that, to establish harmony within the human body (or even the human soul), one must be attuned to the harmony of the world at large. One hundred years before Copernicus received his medical education in Italy, Geoffrey Chaucer's narrator in the *Canterbury Tales* approved the credentials of a doctor, "for he was grounded in astronomye."

As Rheticus observed, studying medicine had important practical consequences for Copernicus's way and pace of life. By the 1530s his patient list read like a who's who of all Prussia. Copernicus treated Maurice Ferber, the second successor to his uncle Lucas in the episcopal palace in Heilsberg, from 1531 until Ferber's death in 1537. This bishop's successor, John Dantiscus (later Copernicus's persecutor), fell seriously ill early in 1538, but with Copernicus in attendance he made a full recovery. Subsequently, in mid-April, Dantiscus wrote to Tiedemann Giese in Kulm, stating: "I feel better, as the doctor, our respected and common friend, will tell you in greater detail. His gentle manner and conversation, and his advice, as soon as I took it, were a cure for me."[24]

Copernicus's reputation as a doctor extended beyond Varmia. In April of 1541 Duke Albrecht of Prussia personally wrote Copernicus an urgent letter asking him to attend a beloved counselor who had fallen ill. Copernicus at the advanced age of sixty-eight traveled to Königsberg with Rheticus and spent three weeks caring for the patient. An earnest correspondence then took place between Copernicus and the duke over the following two months, and the duke also wrote a pious letter to the Frauenburg chapter praising the "God-given skill" of his "dearly beloved honorable and learned Nicolaus Copernicus, doctor of medicine, etc., [the chapter's] colleague and dear, friendly older brother."[25]

Such was the range of acquaintances and skills—and such was the mind—that inspired Rheticus the young mathematician from Wittenberg to write a life of the astronomer in addition to championing his cosmology. As for Copernicus himself, he had now beyond all expectation met one person eager to grasp his astronomical life's work. A new world was dawning, and Copernicus was no longer alone in it.

CHAPTER 4

WHAT RHETICUS KNEW

Few details survive of the first meetings between Rheticus and Copernicus in the spring of 1539. But in addition to the letters of introduction he carried with him from important figures such as Schöner, Rheticus offered Copernicus a gift of six fine books, including an edition of Ptolemy's *Almagest,* plus three titles published by Petreius in Nuremberg: a book by Peter Apian (Rheticus's new acquaintance in Ingolstadt), Witelo's *On Perspective,* and Regiomontanus's *On Triangles,* edited by Schöner himself. These beautifully produced volumes would have been extremely valuable in practical terms, especially to an isolated scholar in Frauenburg. More strategically, they would have shown Copernicus how professionally and in what prestigious company his own work might eventually be published, should he be willing.

The books would have helped Rheticus make a good first impression as a young man of erudition and exceptional contacts—even if they did not eliminate the need to establish his own credibility. He still had to speak for himself and hope that Copernicus would receive him with understanding and generosity.* Between Copernicus, sixty-six,

---

*The quality of Copernicus's welcome can be inferred from Rheticus's own reception of his student Valentin Otto in the parallel meeting between young scholar and old scholar three and a half decades later. "He received me most warmly," Otto would write of Rheticus—who was consciously doing unto another as had been done unto him so many years earlier in Frauenburg (preface to *Opus palatinum,* 1596).

and Rheticus, twenty-five, there quickly grew a cordial teacher-student bond, the more remarkable given that Copernicus had never before had a student, and would never have one again.

Upon his arrival, Rheticus already had some notion of the radical new cosmology which Copernicus had been working on for many years. The richest rumors of this theory had been conveyed to him during his visit to Nuremberg, and it was to the Nuremberger Schöner that he would address his *First Account*. There is a good chance Schöner first learned of the new theory by means of an unpublished manuscript that Copernicus had distributed almost three decades earlier in hopes of evoking response from serious astronomers. This first announcement of heliocentrism—of the sun-centered astronomical system—came to be known as *Commentariolus*, or "Little Commentary," though a coy announcement it was: untitled, unsigned, and circulated privately. Yet from its opening page it exposed its first—and few—readers to a counterintuitive, even shocking, new way of conceiving the universe. This document offers the most direct and convenient survey of what anyone could have known of Copernican cosmology prior to 1540 when Rheticus published his own *First Account* of it.

Copernicus began the *Commentariolus* with the briefest of historical summaries, emphasizing two ancient assumptions (which he never challenged) concerning (1) the regularity and (2) the spherical nature of heavenly motion. "Our ancestors assumed . . . a large number of celestial spheres . . . to explain the apparent motion of the planets by the principle of regularity. For they thought it altogether absurd that a heavenly body should not always move with uniform velocity in a perfect circle."[1] These assumptions of uniformity and circularity were in effect two Aristotelian "demands" at issue in Ptolemaic astronomy's long struggle to "save the appearances."

Copernicus, like Rheticus in his early publications, consistently referred to Ptolemy with great respect, an attitude that appears to have been genuine as well as politic. Moreover, in the *Commentariolus,* Copernicus characteristically adopted an almost flat, "lab report" style. Although he opened it with a slightly caustic reference to an "absurd" hypothesis, his prose immediately turned decorously cool and concessive. Ptolemy's "planetary theories," Copernicus admitted, are "consistent with the numerical data." And yet, he regretted, they "present no small difficulty."

> For these theories were not adequate unless certain equants were also conceived; it then appeared that a planet moved with uniform velocity neither on its deferent [the "circle" that carries it around] nor about the center of its epicycle. Hence a system of this sort seemed neither sufficiently absolute nor sufficiently pleasing to the mind.
>
> Having become aware of these defects, I often considered whether there could perhaps be found a more reasonable arrangement of circles, from which every apparent inequality would be derived and in which everything would move uniformly about its proper center, as the rule of absolute motion requires. After I had addressed myself to this very difficult and almost insoluble problem, the suggestion at length came to me how it could be solved with fewer and much simpler constructions than were formerly used, if some assumptions . . . were granted me.[2]

The equant, as Rheticus knew, was a device invented to rescue and preserve uniform celestial motion but ended up proving quite arbitrary (the opposite of absolute) because the uniformity it produced was merely angular and not "proper." The equant had its purely numerical uses, but scientifically, philosophically, it was problematic. Copernicus called this and the other contrivances of Ptolemaic astronomy merely defects. But after what he hinted was a great deal of hard work, the solution came to him, a solution more reasonable and, as he famously put it, more "pleasing to the mind."

This preamble to the *Commentariolus* takes up only a page, after which appears Copernicus's list of seven postulates or anxioms—a list that presents the Copernican universe in a nutshell. The seven Copernican axioms are as follows:

1. There is no one center of all the celestial circles or spheres.

Though it sounds startling, this is actually the least radical of the seven points. The Ptolemaic theory of eccentrics had already in effect conceded that different circles had different centers. And neither Copernicus himself nor Rheticus ever doubted that the heavenly motions must be circular. (Only with Kepler would this assumption be challenged.)

2. The center of the earth is not the center of the universe, but only of gravity and of the lunar sphere.

Given that the words *geocentric* and *Ptolemaic* are today often popularly treated as equivalent, it may appear a little surprising that Ptolemaic astronomy too had already, in its doctrine of eccentrics, virtually conceded the first part of this Copernican axiom. But with the mention of *gravity* (which in a sixteenth-century context might better be translated merely "heaviness") whole new possibilities open up, possibilities that came to fruition a century and a half after Copernicus in the work of Isaac Newton.

3. All the spheres revolve about the sun as their mid-point, and therefore the sun is near the center of the universe.

This astonishing claim, the signature assertion of the new cosmology, is slightly more vague than it sounds in some modern translations, for Copernicus too retained a version of eccentrics. Thus, in his view, the sun is merely very near (*circa*) the center of the universe. His universe, in other words, was actually only *almost* heliocentric.

**4. The ratio of the earth's distance from the sun to the height of
the firmament is so much smaller than the ratio of the earth's
radius to its distance from the sun that the distance from the earth
to the sun is imperceptible in comparison with the height of the
firmament.**

Among the seven axioms, this is probably the most shocking in its consequences, even if on casual first reading it sounds merely abstract. According to Ptolemy, the universe (the sphere of the fixed stars, or the
"firmament") is so large "that the earth has the ratio of a point to the
heavens."[3] This conclusion is supported empirically by the fact that
the plane of earth's horizon always appears to bisect the universe exactly, whereas it ought to be off-center by the distance between the center of the earth and the observer's position on earth's surface (i.e., by
one earth radius). From the absence of any discernible difference between the cosmic location of earth's center and that of a point on its
surface, Ptolemy logically concluded that the universe is *immensely* (literally, *immeasurably*) vaster than the earth; and conversely, in relation
to the universe the earth is immensely small.

However, Copernicus knew that (in his system) an observer's stand-
point on the circle of earth's pathway around the sun *likewise* appears to
bisect the sphere of the fixed stars. Accordingly, as indicated in axiom 4,
the height of the firmament—the distance from the earth to the sphere
of the fixed stars—is *so* immense that the earth-sun distance is also as
nothing by comparison. Of course, it was well known that the earth-
sun distance is actually very large indeed. How enormous, then, must
be that immensity which makes "very large" appear as nothing. Copernicus's axiom 4, in short, established a radical new benchmark for the
immensity of the universe.

Granted a few modest assumptions, one can calculate approximately
how much more immense Copernicus's cosmos must be than Ptolemy's.
For Ptolemy, the universe was immensely large in relation to the earth;

but for Copernicus, it was immensely large in ralation to *the size of earth's circular pathway.* In the astronomy of Copernicus's day, the distance from the earth to the sun was believed by astronomers to be 1,210 times greater than the earth's radius.[4] This implies that if earth were to circle about the sun, the diameter of that orbit would be 1,210 times greater than the diameter of the earth. Since size (volume) increases in a cubic proportion to linear distance, it may logically be concluded that $1,210^3$ describes how many times vaster Copernicus's cosmos was than Ptolemy's. And $1,210^3$ (i.e., 1,210 times 1,210 times 1,210) equals 1,771,561,000. It can therefore be conservatively stated that Copernicus's fourth axiom entailed a universe well over a billion times larger than the conventional value accepted by most of his predecessors and contemporaries.*

It was this stunning conclusion, perhaps as much as the preposterous claims regarding the earth's motion, that initially made the Copernican system so hard to accept.

5. Whatever motion appears in the firmament arises not from any motion of the firmament, but from the earth's motion. The earth together with its accompanying elements performs a complete rotation on its fixed poles in a daily motion, while the firmament and highest heaven abide unchanged.

This axiom, which marks what is usually called the rotation of the earth, dealt a coup de grâce to one of the hardest-to-imagine tenets of Ptolemaic astronomy: the claim that the entire immense stellar sphere makes a full revolution every twenty-four hours.

---

*According to Archimedes' work known as *The Sand Reckoner,* speculations concerning the immensity of a heliocentric universe were also engaged in by the ancient Greek thinker Aristarchus of Samos (ca. 310 B.C.–ca. 230 B.C.). See *The Book of the Cosmos: Imaging the Universe from Heraclitus to Hawking,* edited by Dennis Danielson (Cambridge, Mass.: Perseus, 2000), chap. 7.

6. What appear to us as motions of the sun arise not from its
motions but from the motion of the earth and our sphere, with which
we revolve about the sun like any other planet. The earth has, then,
more than one motion.

One of the most familiar Copernican assertions, this axiom opened up
a whole new way of "reading" the Book of Nature, according to which
the state and motion of the perceiving subject (i.e., the observer) is really what accounts for the apparent state or motion of the object of
perception (in this case, the sun). Simply, this axiom established the
second of earth's motions—its revolution or orbit about the sun—and
stated scientifically, for the first time in the history of the world, that
earth is a planet.

7. The apparent retrograde and direct motion of the planets arises
not from their motion but from the earth's. The motion of the earth
alone, therefore, suffices to explain so many apparent inequalities in
the heavens.

Continuing the scientific explanation of "subjectivity" from axiom 6,
this axiom offered to solve at a stroke one of the great embarrassments of
Ptolemaic astronomy: that despite its strenuous efforts to save the appearances, it could not describe the heavens in a manner that rescued
their regularity and uniformity of motion. Regardless of how radical this
list of axioms was, Copernicus ended it by reminding his readers that he
was attempting to conceive of a system that would achieve what astronomers had been seeking all along: an absence of real inequalities in
the heavens. What he longed for instead was a comprehensive, harmonious vision of a truly unified universe.

The *Commentariolus*, copied in manuscript and sent out to an unknown
number of scholars for their reactions, was similar to a preprint today
that a scientist posts in hopes of receiving some helpful, expert response

before the paper is submitted for wider publication. The manuscript copies of the *Commentariolus*, however, met with what one science historian calls a "tantalizing silence": "We have no idea to whom Copernicus may have sent the original copies . . . We have no idea whether any of the recipients replied to him, although some of them must have, or what they may have said. We have recovered no mention of it during the decades that followed."[5]

The silence is not so surprising. The substance of the *Commentariolus*, summarized in the seven axioms, spelled such a radical departure from millennia-old assumptions and habits of intellect that it required a rare kind of mind and motivation on the part of one who would both comprehend and accept it.

CHAPTER 5

COPERNICAN SUNRISE

Despite the commonly held belief that Copernicus was somehow too timid to publish his theory, he had numerous other reasons for leaving his work unfinished for so many years. In the first place, it is hard to muster the effort required to publish a book unless one believes there will be an audience for it. And his duties as an administrator in Varmia during decades of economic turmoil and military conflict further distracted him from his astronomical quest.

In 1524 Copernicus had said he intended to set forth his views "concerning the motion of the sphere of the fixed stars."[1] Given the demands of his many political, administrative, and medical responsibilities, however, added to the lack of any response to his *Commentariolus*, it is not surprising his good intentions went unfulfilled for almost another two decades. Yet Copernicus's love of astronomy never died. He only needed someone else to share it with him—and help share his vision effectively with the wider world.

Within days of Rheticus's arrival in Frauenburg, an apparently rejuvenated Copernicus was hard at work teaching his new student about his astounding theory of the universe; and Rheticus, in a state that must have bordered on intellectual vertigo, was struggling to grasp the immensity of what he was reading and hearing.

Many hours were taken up with celestial observations conducted

from the *pavimentum*—a sort of level patio—that Copernicus had prepared beyond the fortifications that surrounded Frauenburg Cathedral,
between the northwest tower (now called Copernicus's Tower) and his
house (*extra muros*—"outside the walls") that stood some hundreds of
feet farther to the west. From the *pavimentum* Copernicus and Rheticus
had an unobscured view north and south along the meridian, so that
with their backs to the sea they could chart—with as much precision as
the naked eye afforded—the courses of the wandering stars.

On top of his own strenuous wanderings over the previous nine
months, all this work threatened to leave Rheticus physically and men-
tally exhausted. By his own later report, his health soon began to fail,
and in response his new teacher the doctor prescribed a period of rest
and relaxation. Copernicus too welcomed the chance for a break, and
both men repaired to the episcopal residence of Copernicus's best
friend, Bishop Tiedemann Giese, in Löbau, some seventy miles south
of Frauenburg.

The young Protestant from Wittenberg and the elderly Catholic
bishop took an instant liking to each other, strengthened by their
shared affection for Copernicus and their mutual commitment to cultivating knowledge—to furthering what Rheticus along with other humanists called the republic of letters.

During the several weeks of leisure they spent together in midsummer of 1539, Giese and Rheticus encouraged Copernicus to reveal his
theories to the world. They urged him not only to offer practical astronomical tables and predictions, but also, "in imitation of Ptolemy," to
declare boldly "the system or theory and the foundations and proofs
upon which [these] relied"—even if such theories "contradict our
senses" and are "diametrically opposed to the hypotheses of the ancients." On the other hand, Rheticus would note, Giese's advice to
Copernicus was soberly realistic in anticipating that a view as radical as
his would face the tendency "of scholars everywhere to hold fast to their
own principles passionately and insistently."[2]

Having heard the words of Giese and seen his teacher's lingering hesitation, Rheticus hatched a most decisive scheme. He would tell a story, offer a *narratio,* that would double as advertisement and trial balloon for Copernicus's theory.

Back in Frauenburg, a revived Rheticus set to work again, this time on an accessible introduction to the very theory he himself was still struggling to grasp. The fastest way to learn a subject is to put oneself in a position of having to teach it, and by September 23 Rheticus had shown himself to be a consummate teacher as well as learner—completing the work that would become the world's first Copernican publication. The draft was soon approved by both Giese and Copernicus, and Rheticus left for Danzig in search of a publisher.

Copernicus could not have wished for a more gifted agent and impresario than Rheticus. Not only had this sociable young dynamo rekindled Copernicus's interest in his own work and prepared a brilliant précis of it; he was also now in the process of displaying his exceptional native talent for diplomacy and personal networking. The manuscript of the *First Account* (*Narratio prima*) that he was carrying with him to Danzig was formally addressed to Johann Schöner in Nuremberg. Like any perceptive author, however, Rheticus understood that a crucial slice of his audience consisted of those who would read his work prior to its publication. This awareness helps explain a highly peculiar composition that he tacked onto the end of his *First Account.*

The essay, titled "In Praise of Prussia" (*Encomium Borussiae*),[3] was written in an extravagantly florid style. Nobody reading it can doubt that Rheticus was euphoric.

Rheticus began it with a strained if well-meaning analogy between the island of Rhodes, "dearly beloved spouse of the sun," and Prussia, where the sun offers its blessings "by lingering above the horizon"—a reference to the long summer days enjoyed in northern latitudes. After praising the natural endowments of Prussia, Rheticus lauded its ruler,

Duke Albrecht, "patron of all the learned and renowned men of our time," and rhapsodized over the city of Danzig, "metropolis of Prussia, eminent for the wisdom and dignity of its Council, for the wealth and splendor of its renascent literature." Rheticus ended the first part of his *Encomium* with a flattering, idealized picture of the educational culture of Prussia: Just as Aristippus and his companions, shipwrecked on Rhodes, gathered hope when they "noticed certain geometrical diagrams on the beach"—evidence of the presence of "educated and humane men"—so when Rheticus entered the homes of the hospitable Prussians (so he informed Schöner), he would see "geometrical diagrams at the very threshold" or else find "geometry present in their minds."

Rheticus knew that these words, before they ever reached Schöner in Nuremberg, would be read by those inhabitants of Prussia upon whose goodness and kindness he was dependent for getting his work published. Rheticus identified first Tiedemann Giese and then "the second of [his] patrons," "the esteemed and energetic John of Werden, ... mayor of the famous city of Danzig." Rheticus's effusive picture of the mayor combined heroism (he was like Homer's Achilles) and cultural refinement, for he was a lover of music, and "by its sweet harmony he refreshes and inspires his spirit to undergo and to endure the burdens of office."

The most obvious and predictable result of this compelling flattery was that in Danzig Rheticus readily received the mayor's hospitality and quickly found a publisher. Equally important, Rheticus's need to appeal to John of Werden's character and tastes may have helped him crystallize in the *First Account* what would become one of the dominant metaphors of Copernican cosmology: musical harmony. The mayor, as Rheticus wrote, cherished music (*musicam colit*). And at the conclusion of the *Encomium* Rheticus accordingly meditated on the analogy— both parts of which were relevant to Werden—between music and the government of a state: "The soul of a heroic man . . . is called a harmony.

Hence we might correctly call those states happy whose rulers have harmonious souls."

As both Copernicus and Rheticus knew, a fine piece of music, a good man, a well governed state, and a cosmos created and governed by a good God—all must embody harmony.

Rheticus spent much of his life looking for love, for beauty, for belonging. But what he sought above all was a hearing for Copernicus and his ideas, and in fact Rheticus's own name appeared nowhere in the extended title of the monumental little work that ought to have made him famous.* The title did explicitly name both Schöner and Copernicus, however, and its many superlatives echoed its author's enthusiasm.

TO THE MOST ILLUSTRIOUS GENTLEMAN
MR. JOHANN SCHÖNER, CONCERNING
THE BOOKS OF THE REVOLUTIONS
Of the most learned Gentleman and
Most distinguished Mathematician,
The revered Doctor Mr. Nicolaus
Copernicus of Toruń, Canon of
Varmia, By a certain youth
Most zealous for
Mathematics—
A FIRST ACCOUNT.

Below this title—set like an equilateral triangle, base upward but balanced on a plinth—stands a Greek proverb by Alcinous, one of Rheticus's favorites: "Free in mind must be the one who longs for wisdom."

Despite the fact that he, not Copernicus, broke the news of

---

*Perhaps some fame is beginning to attach itself to Rheticus's work. In 2004 the Linda Hall Library of Science, Engineering & Technology in Kansas City paid $1.5 million for a 1540 first edition of *Narratio prima (First Account)*.

Title page of the world's premier Copernican astronomical publication: Rheticus's First Account *(*Narratio prima, *1540).*

Copernicanism to the world, and that this *First Account* would go through four editions in the sixteenth century, twice as many as *The Revolutions* itself, and that it would probably also reach more readers than the larger work, Rheticus on the title page humbly referred to himself—accurately if anonymously—merely as someone characterized by a love of mathematics. His sole interest was in reflecting, not deflecting, the light that shone from the mind of his teacher.

The first third of the *Account* grapples in a highly technical way with the problem known as the "precession of the equinoxes." The earth's axis, rather like that of a spinning top, is ever so slowly wobbling westward (the opposite direction from earth's rotation) around the pole of the ecliptic. This means that the equinoxes—the starting points of spring and autumn—are gradually slipping backward relative to the constellations and all the other stars as well. The full cycle of the precessional wobble takes approximately twenty-six thousand years to complete.

The phenomenon of precession, which astronomers had been aware of since Hipparchus in the second century BC, was never fully explained by Rheticus or Copernicus. However, Rheticus understood that precession was virtually impossible to explain if one assumed a central, immobile earth. Thus his slow, painstaking treatment of the problem in effect aimed at softening up the anticipated opposition and preparing the reader to be introduced to a more promising theory. And yet, given the centuries of respect paid to Ptolemy, the true opposition must not appear as opposition. Rheticus went to great lengths to honor Ptolemy and assert that Copernicus was building upon, not tearing down, the work of that great astronomer. Copernicus merely had a longer, fuller sample of calculations to guide him.

> Ptolemy's tireless diligence in calculating, his almost superhuman accuracy in observing, his truly divine procedure in examining and investigating all the motions and appearances ... cannot be sufficiently admired ... However, a burden greater than Ptolemy's confronts my teacher. For he must arrange in a certain and consistent scheme or harmony the series and order of all the motions and appearances, marshalled ... by the observations of 2,000 years ... Ptolemy was able to harmonize satisfactorily most of the hypotheses of the ancients ... [And] he quite rightly and wisely—a praiseworthy action—selected those hypotheses which seemed to be in better agreement with reason and our senses ... Nevertheless, the observations of all scholars and heaven itself and mathematical reasoning convince us that Ptolemy's

hypotheses and those commonly accepted do not suffice to establish the *perpetual and consistent connection and harmony of celestial phenomena* and to formulate that harmony in tables and rules. It was therefore necessary for my teacher to devise new hypotheses, by the assumption of which he might geometrically and arithmetically deduce . . . systems of motion like those which the ancients and Ptolemy, raised on high, once perceived "with the divine eye of the soul." . . . Surely students hereafter will see the value of Ptolemy and the other ancient writers, so that they will recall these men . . . and restore them, like returned exiles, to their ancient place of honor. The poet says: "No one desires the unknown."

Rheticus's account was brilliantly inconsistent. He and Copernicus were about to overthrow Ptolemy, yet the early pages of his *First Account* offered a picture of Copernicus as Ptolemy's rescuer. Then, as the argument reached its climax, he pulled out the rhetorical stops, employing figurative language ("like returned exiles") and quotation from "the poet"—from Ovid's *The Art of Love*. Still, at the core of his case for Copernicus was the demand that the universe be viewed as a whole (by a mind raised on high) and conceived of as involving "consistent connection and harmony."

From this moving plea for a harmonious universe, Rheticus moved decisively to the signature if counterintuitive idea of the new Copernican cosmology: a moving earth. As he triumphantly informed Schöner, Copernicus would liberate us from equants. Furthermore, declared Rheticus, the appearances of the planets—their varying distances from earth and their looping courses—could be explained "by a regular motion of the spherical earth; that is, by having the sun occupy the center of the universe, while the earth revolves instead of the sun." That pathway of the earth Copernicus named "the great circle" (*orbis magnus*), and indeed "there is something divine in the circumstance that a sure understanding of celestial phenomena must depend on the regular and uniform motions of the terrestrial globe alone."

In the space of a few lines Rheticus swept away two of the most troubling Ptolemaic devices (equants and epicycles) and transformed a third. What had thus far been the circle of the sun was to be boldly appropriated by the earth. And the result was the greatest astronomical prize of them all: a unified explanation of the phenomena.

Both Copernicus and Rheticus saw "the great circle"—what is now known as earth's annual orbit—as great not only in size but also in magnificence and explanatory power. It alone was the key to unifying the varying motions of the planets. It harmonized beautifully with the fact that the "inferior" planets have proportionately shorter periods of revolution than does the earth, while the "superior" planets have proportionately longer ones.

As Kepler would later suggest, earth's circle was great also in the sense of being privileged and even (in a new sense) central, since it could be viewed as tracing a path midway between the sun, Mercury, and Venus on the one hand and the three superior planets—Mars, Jupiter, and Saturn—on the other.[4] As Rheticus commented in the *First Account,* "Almost alone it makes us share in the laws of the celestial state."[5] Most important, the great circle's multiple functions point in turn to the abundant wisdom of God. "So wise is our Maker," declared Rheticus, quoting Galen, "that each of his works has not one use, but two or three or often more."[6]

For astronomy and cosmology the great circle established a "common measure"—a phrase integral to Rheticus's dominant metaphor of harmony, which functioned as a preempirical assumption governing his theorizing about the universe. With a gracious bow in the direction of Ptolemy, Rheticus summed up his musical analogy in the following carefully crafted words.

> My teacher was especially influenced by the realization that the chief cause of all the uncertainty in astronomy was that the masters of this science (no offense is intended to divine Ptolemy, the father of astronomy) fashioned their theories and devices for correcting the motion of the

heavenly bodies with too little regard for the rule which reminds us that the order and motions of the heavenly spheres agree in an absolute system. We fully grant these distinguished men their due honor, as we should. Nevertheless, we should have wished them, in establishing the harmony of the motions, to imitate the musicians who, when one string has either tightened or loosened, with great care and skill regulate and adjust the tones of all the other strings, until all together produce the desired harmony, and no dissonance is heard in any.[7]

Only twenty-five years old, Rheticus longed to be credible and convincing in the eyes of Schöner and other learned men. In this attractive ancient metaphor of the world as a finely tuned instrument, he had struck just the right persuasive balance between piety toward his forebears and passion for what he so fervently hoped would be the cosmology of the future.

One remarkable element in the birth of Copernicanism was how poetry served as a vehicle for scientific discovery and exposition. Popularly, poetry and science are today seldom treated as close kin, or even close friends. However, numerous eminent minds who have studied the nature of poetry and science recognize a genuine affinity, most notably in the structures displayed by poetic metaphors and scientific models.[8] One of the most common and well-known poetic devices is the simile. Yet similes also have an important if less conspicuous function outside of formal poetry. Often a person's most elementary request for information about an unknown thing, event, or experience is "What was it like?" In this basic way, simile or poetic analogy is essential to the acquisition of new knowledge.*

One of the most famous passages Copernicus ever penned appeared

---

*Although Rheticus took poetry and figurative language seriously, not all the poetry he enjoyed was serious. One of his surviving poems developed a farcical analogy between the "progress" of too much Breslau beer through the body of a drinker on the one hand and the succession of the constellations of the Zodiac on the other. Titled "De XII. Signis Zodiac ac Cerevisia Vratislaviensi apud Silesios" ("The Twelve Signs

in book 1, chapter 10, of *The Revolutions*. In an outburst of figurative, literary language, Copernicus wrote:

> And behold, in the midst of all resides the sun. For who, in this most beautiful temple, would set this lamp in another or a better place, whence to illuminate all things at once? For aptly indeed do some call him the lantern—and others the mind or the ruler—of the universe. Hermes Trismegistus calls him the visible god, and Sophocles' Electra "the beholder" of all things. Truly indeed does the sun, as if seated upon a royal throne, govern his family of planets as they circle about him . . . Thus we discover in this orderly arrangement the marvelous symmetry of the universe and a firm harmonious connection between the motion and the size of the spheres such as can be discerned by no other means.[9]

Copernicus thus honored the sun by calling it lamp, lantern, mind, ruler, visible god, beholder of all things, and royal governor. And in so doing he attempted to avoid or at least blunt what could have been one of the most serious criticisms of his theory.

Copernicus knew that his ideas could be seen by his contemporaries as removing the sun from the heavens and depositing it in the lowest, least exalted place: in the dead center of the universe.[10] According to prevailing Aristotelian natural philosophy, the cosmic center was where heavy—and by extension ignoble—things collected. As one influential philosopher had expressed it in the previous century, the center represented the "the excrementary and filthy parts of the lower world."[11]

Anticipating the charge that he was dishonoring the sun, Copernicus deftly reconstrued the center as a place of honor and of effective, efficient government. Using a poetic play on words, he imagined it as a throne (*solium*) fit for the sun (*sol*). With the sun in the center, the path-

of the Zodiac and the Silesian Beer of Breslau"), it was published by his friend Caspar Brusch in his *Sylvarum* (Leipzig, 1544). A transcription and a rough translation appear in Jesse Kraai, "Rheticus' Heliocentric Providence: A Study Concerning the Astrology, Astronomy of the Sixteenth Century." (Ph.D. diss., University of Heidelberg, 2001), pp. 274–281. Available online at http://www.ub_uni-heidelberg.de/archiv/3254.

ways of the planets at last made sense; they displayed marvelous symmetry and harmonious connection; they exemplified the sort of beauty that something called a *cosmos* ought to embody: The sun's and the planets' astronomical truth reflected poetic truth and, in a profound sense, also poetic justice and decorum.

In the *First Account* Rheticus used the same governmental analogy to support the Copernican claim that the sun is at rest, that it does not move about like a planet. "My teacher . . . is aware that in human affairs the emperor need not himself hurry from city to city in order to perform the duty imposed on him by God."[12] The centrality and immobility of the sun in Copernicus's system were therefore perfectly consistent with—and were indeed essential to—the sun's dignity and its proper governance of the planets. As Rheticus declared, "While we were unable from our common theories even to infer this rule by the sun in the realm of nature, we ignored most of the ancient encomia of the sun as if they were merely poetry."[13]

But the encomia—the songs of praise—were nothing less than poetry, and in fact turned out to be not distractions from reality but signposts pointing toward it. The similes and metaphors of the sun as ruler were themselves clues to the real physical structure of the sun and the universe. Copernicus's system did imply that the sun occupies literally the lowest possible cosmic location—that, in Rheticus's words, it "has descended to the center of the universe."[14] Nevertheless, the fitness, beauty, and glory of the sun in this reconfigured cosmos more than compensated for that lowering. In the *First Account* Rheticus expressed that glory in explicitly poetic form.

> Thus God stationed in the very midst of this theater his governor of
> nature, king of the entire universe, conspicuous by its divine splendor, the sun
>> To whose rhythm the gods move, and the world
>> Receives its laws and keeps the pacts ordained.[15]

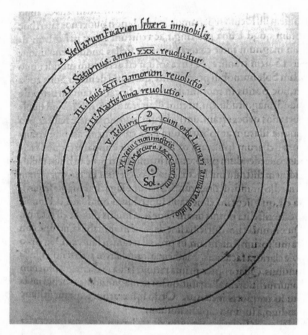

Copernicus's famous diagram of the spheres (here from The Revolutions, *1566), with the periods of the planetary revolutions harmoniously increasing in relation to the distance of each one from the central sun.*

Having exchanged places with the sun, the earth was now risen in the world, promoted to the status of a "star," and "move[d] among the planets as one of them."[16] As Galileo would later declare, earth was no longer "excluded from the dance of the stars."[17] Rheticus and Copernicus together achieved the monumental feat of raising the status of both earth and sun within the same bold cosmology.

CHAPTER 6

GOD'S GEOMETRY IN HEAVEN AND ON EARTH

With the appearance of the *First Account* early in 1540, Rheticus and his supporters launched an energetic campaign on its behalf. Even before its full publication in March of that year, Andreas Aurifaber, a university friend of Rheticus's then staying in Danzig, sent the book's first three printed sheets to Melanchthon back in Wittenberg.[1] The gesture was likely intended to make Melanchthon feel included in Rheticus's work—and to feel that Rheticus was doing more in Varmia than merely neglecting his academic duties in Wittenberg.

On April 23, 1540, Tiedemann Giese also sent a copy of the newly published work to Duke Albrecht in Königsberg. Beyond any hint of sectarianism, the Catholic bishop included a short, formal cover letter in German informing the Protestant duke of the efforts that a Protestant academic was making to further the work of the Catholic astronomer. In return, Giese requested, "May Your Princely Eminence look graciously upon this highly learned guest on account of his great knowledge and skill, and grant him your gracious protection."[2] Even if the duke had no scientific influence on how Copernicanism was received, he did respond graciously—and crucially—to the request that he show personal favor to Rheticus.

Other promotional copies of Rheticus's little book had an even greater impact. The copy received by its official addressee, Johann

Schöner, and then handed on to his fellow Nuremberger Petreius, gar-
nered an eager response: "I consider it a glorious treasure," the printer
wrote, "if some day through your urgings [Copernicus's] observations
will be imparted to us."[3] Now Rheticus could show Copernicus docu-
mentary evidence proving there was serious interest in his work in the
wider world of learning.

When the famous Dutch cosmographer Gemma Frisius received a
copy of the *First Account*, he wrote to Bishop Dantiscus, declaring: "I
am filled with desire to see this business brought to fruition. And
everywhere there are more than a few erudite men whose minds desire
it no less than I do."[4] The entrepreneurial Rheticus, by composing,
publishing, and distributing the *First Account*, was able to set before his
teacher compelling proof that *The Revolutions* would have avid readers
and a willing, well-placed publisher too.

Rheticus sent a further gift copy of the *First Account* to his old mentor Achilles Gasser in Lindau, who in turn passed it on, accompanied
by a detailed letter, to his friend Georg Vögelin in Constance.[5] The
new cosmology, Gasser wrote, "will amaze and utterly astonish you"
and "may even appear . . . to be (as the monks would say) heretical."
Gasser also hoped that Vögelin would contribute to "a stream of requests" being sent to Copernicus requesting that he divulge "his whole
work . . . by means of the persistence, effort, and tireless diligence of
[Gasser's] friend [Rheticus]."[6] Vögelin responded with ten lines of
verse reprising the theme of astonishment—as well as possible
hostility—with which people would respond to Copernicus's theories.

> *Unknown to ages past—to minds today*
> *Astonishing—the contents of this book:*
> *For here the ranks of stars receive new orders,*
> *And earth, once thought to stand, now hastens round.*
> *Praise be to ancient skill for arts invented,*
> *But no less praise to achievements newly born.*
> *An envious, not a cultivated, mind*

Dreads or despises teachings that are new.
Yet should ill will forbid wide approbation,
Suffice it that this work delight the learned.[7]

Gasser's letter and Vögelin's poem were printed along with the second
edition of the *First Account* in Basel in 1541 and formed the earliest published response to Rheticus's announcement of the Copernican universe.

* * *

DE LIBRIS REVO.
LVTIONVM ERVDITISSI.
MI VIRI, ET MATHEMATICI
excellentiſſ. reuerendi D. Doctoris
Nicolai Copernici Torunnæi Cano
nici Vuarmacienſis, Narratio Prima
ad clariſſ. Virum D. Ioan. Schone-
rum, per M. Georgium Ioachi-
mum Rheticum, unà cum
Encomio Boruſſiæ
ſcripta.

ALCINOVS.

Δᾶ δ᾽ ἐλιυθριου ἄναι τῇ γνωμῇ τὸρ
μάλλοντα φιλοσοφᾶν.

GEORGIVS VOGELINVS ME-
dicus Lectori.

Antiquis ignota Viris, mirandaqꝫ noſtri
    Temporis ingenijs iſte Libellus habet.
Nam ratione noua ſtellarum quæritur ordo,
    Terraqꝫ iam currit, credita ſtare prius.
Artibus inuentis celebris ſit docta Vetuſtas,
    Ne modo laus ſtudijs deſit, honorꝙ nouis.
Non hoc iudicium metuunt, limamqꝫ periti
    Ingenij, ſolus liuor obeſſe poteſt.
At ualeat liuor, paucis ettam iſta probentur·
    Sufficiet, doctis ſi placuere Viris.

B A S I L E AE.

Title page of the First Account, *second edition (1541),*
displaying Georg Vögelin's poem.

If Rheticus's main task in Varmia was to announce Copernicus's work to the world, by early 1540 he had already essentially done so. Rather than returning to Wittenberg, however, he felt personally driven to stay by the side of Copernicus in order to see his teacher's larger work, *The Revolutions*, through to completion. Despite the awkwardness of his taking yet more time away from academic duties, he had found in Prussia work and people he adored—and, for now, could not bear to leave. Although he had written and published the *First Account*, he was only beginning to grasp the ideas, and the mission, that lay before him.

That mission would not end with the publication of *The Revolutions* but would include the preparation of mathematical tools for the verification and extension of what might be called the Copernican revelation. Part of the essence of this revelation was that the universe, as Rheticus had asserted in the *First Account*, is a harmonious unity. Contrary to the not-yet-discarded image of the universe, there was no upper story or lower story. It was all a single structure, an undivided kingdom, with God's deputy the sun in the center governing the stars and the planets. Moreover, if the whole universe exhibited order and measure, by logical implication it could be measured, and the same standard of measurement would apply throughout the kingdom. That standard was geometry.

Although Galileo often receives primary credit for applying geometry (literally "earth-measure") to celestial space,[8] the justification for doing so was asserted by Rheticus. For him it was an integral part of the Copernican revelation. As he would later state explicitly, mathematicians are to teach "God's geometry in heaven and on earth,"* and "the heavens speak to us by means of astronomy, the earth by means of geography."[9]

---

*This gentle echo of the Lord's Prayer neatly underlines Rheticus's vision of a unified kingdom, with geometry itself as a manifestation of God's will.

Detail from the title page of Peter Apian's Introduction to Geography *(1533) illustrates the importance of triangles to astronomical as well as earthly measurement.*

In applying geometry to geography as well as astronomy, Rheticus continued to follow in Copernicus's footsteps. Years earlier, Copernicus had prepared a preliminary map of Prussia, and Rheticus was keen to extend the theoretical dimension of his teacher's work in this area. His famulus, Heinrich Zell, also happened to have a strong penchant for mapmaking, so it seemed practicable as well as logical, on the heels of the *First Account*, to spend some time pursuing the discipline of geography.

As part of the *First Account*, Rheticus had published the *Encomium*, in which he extolled the glories of the Prussian landscape. Such praise was political as well as geographic; pursuing scientific work with such a down-to-earth focus would afford Rheticus further opportunity for ingratiating himself with Albrecht, duke of Prussia.

For three weeks in Königsberg in the spring of 1541 Rheticus enjoyed the hospitality of the duke while Copernicus was attending Albrecht's ailing counselor. Albrecht had strong personal and religious connections with both Wittenberg and Nuremberg—where his name carried considerable weight on account of his earlier conversion to Lutheranism. In 1540 he had already received from Bishop Giese a fresh copy of Rheticus's *First Account*, so there were numerous mutual interests that Rheticus and the duke could have pursued in conversation. Rheticus worked hard to deepen his contact with the duke, and by August of 1541 he had prepared another offering for Albrecht: a manuscript he called *Chorography*, a title denoting "local geography."

Rheticus's *Chorography* was a short, technical treatise promoting the importance of careful geometric measurement for anyone wanting to make a useful local or regional map—or "chorographical table," as he called it.[10] In this work Rheticus showed how one must begin by establishing the latitude and longitude of one's fixed points, such as cities—which is actually to orient them to the sun and the stars. Rheticus cited the great Nuremberg artist Albrecht Dürer's instructions on how artists can "counterfeit a landscape."* Then, having set one's points and established their coordinates and mutual distances, Rheticus explained, one can start doing what he himself would spend much of the rest of his life doing: calculating triangles.

Rheticus's dream, geographically speaking, was reliable maps drawn to scale, showing "cities, plains, rivers, etc.," and useful for travelers. Such a proposal, not unique to Rheticus, was important enough on its own. But what repeatedly emerges from this little work—written in German, strongly inflected with the dialect of his home province, and

---

*In Rheticus's day *Landschaft* was a new term in German, one that would not make its way into English as *landscape* for another sixty years.

unpublished until three centuries after his death—are its flashes of passion and insight.

For example, Rheticus emphasized the significance of accurate maps and measuring devices for travel on sea as well as land. He had spent some time in conversation with sailors in the port of Danzig—and he was not impressed with what they told him. As the twenty-seven-year-old scientist wrote, not without a touch of self-conscious exasperation: "Many ships sail from Prussia to England and Portugal, and not only do they generally use no measurement of latitude, but they also heed neither chart nor proper compass. The sailors boast that they carry all this know-how in their heads. Which is just fine, as long as things go well. But unfortunately, they often lose that know-how somewhere in their heads and cannot find it when they need it most—so that they with their goods and passengers go nowhere. In my opinion, it would not hurt for these fellows to learn just a bit more about such matters."[11]

The greatest burst of insight and intellectual excitement in the *Chorography* was also the most significant scientifically. Rheticus marveled at the virtues of the magnet, praising its usefulness both nautically and geographically. Yet he also expressed wonder that went beyond pragmatism: The magnet's behavior is a "beautiful spectacle of nature" and shows again "how wonderful is the Lord God in all his works."

It is evident that, here in the final paragraphs of the *Chorography*, Rheticus was on the verge of genuine discovery. Carefully citing the thirteenth-century writer Petrus Peregrinus, to whose work on magnetism he had been introduced by Achilles Gasser, Rheticus commented that the magnet and the heavens share a common property. "If the lodestone is shaped into a sphere, and hung up vertically upon its poles, that is, upon its north and south, . . . then, in keeping with the property with which God has imbued it, it supposedly rotates once in twenty-four hours, just as the sun circles the earth in a day and a night."[12]

Rheticus was mistaken in assuming that magnetic behavior indicates the lodestone's orientation to the heavens. The magnet rotates once in twenty-four hours because the earth does—a fact obscured by Rheticus's oddly obsolete reference to the circling sun. Yet his imaginative picture of a spherical magnet suspended in space and rotating about its axis—later known as a *terrella*, a miniature earth—is suggestive beyond the incorrect details of his exposition. Fifty years later, that is what William Gilbert, under the inspiration of Copernicus, would so decisively declare the earth itself to be: one giant rotating magnet.[13]

Winning the attention and favor of the duke of Prussia was no small achievement. Rheticus had benefited from the preexisting cordial connection between Albrecht and Copernicus, and also Giese. But one of Rheticus's most useful skills was his capacity to recognize—and then seize—big opportunities. Accordingly, with a powerful man on his side, he displayed no shyness in requesting further support. And the duke, to his credit, seemed pleased to encourage the young scientist's ambitious schemes.

Only one day after sending his *Chorography* to Albrecht, Rheticus wrote another letter to accompany the gift of a small instrument for use in measuring the length of days. At the end of the letter he asked for a very particular favor: "that Your Princely Grace should graciously request the Prince-Elector of Saxony and the praiseworthy University of Wittenberg, that I might be granted permission to bring to press the work of my honored teacher." The duke's secretary, Jerome Schurstab, even though he imagined it was Rheticus's own work he wanted to have published, knew exactly the kind of practical endorsement Rheticus needed.

Three days later, on September 1, 1541, Schurstab wrote on the duke's behalf to the prince-elector of Saxony, with a copy to the University of Wittenberg. Rheticus the "honorable and upright" mathematician had spent his time in Prussia pursuing "by divine grace his skill in astronomy" and so bringing "no little luster, fame, and honor not only to Your Eminence but also to the whole university." Moreover, "in recognition of his intelligence, skill, and virtue," Schurstab requested in the duke's name that Rheticus be permitted to publish "his book" at the place of his choosing and—crucially—"without interruption of his salary."[14] Rheticus already hoped to take his teacher's manuscript beyond Wittenberg to Nuremberg, and Schurstab's request supported his rather audacious plan for yet a further extension of his three-year sabbatical.

With political support assured and the first great stage of his scientific mission accomplished, it was time for Rheticus to leave a place he had come to love, and an old man he loved even more. Copernicus had finished *The Revolutions* and entrusted a manuscript copy of his monumental work into Rheticus's hands. With the duke's glowing endorsement, the young entrepreneur was guaranteed a welcome back in Wittenberg. And with Petreius's presses waiting in Nuremberg, Copernicus was guaranteed a publisher. All the pieces were falling into place; Rheticus was about to bring an immense undertaking to the consummation he had longed for.

In that September of 1541, as Rheticus collected his books, papers, and thoughts in preparation for his departure, he had no illusions about his chances of seeing his teacher ever again. Copernicus was already sixty-eight, and Wittenberg and Nuremberg were long journeys from Frauenburg. Rheticus also had many other obligations to repay. This parting would be a final farewell.

Copernicus must have realized this, too. If anything, he was even more keenly aware than Rheticus of how their scientific mission now depended upon the skill and effort of the disciple whose arrival in Frauenburg more than two years earlier had been so startling and so momentous. Both Rheticus and Copernicus knew the story of the great astronomer Georg Peurbach, who on his deathbed eighty years earlier had, fatherlike, bequeathed his legacy and unfinished work upon his student, Regiomontanus, whose *On Triangles* Rheticus brought as a gift to Copernicus in 1539.

It was time for the torch to be passed once more. When Rheticus sketched the scene for King Ferdinand years later, he would liken his teacher to the ancient world's "father of astronomy," calling Copernicus "the Hipparchus of our age," whom "no one has words enough to praise adequately." "Upon my departure that noble old man solemnly charged me to strive to finish that which old age and death would prevent him from completing himself."[15]

CHAPTER 7

COPERNICAN AFTERLIFE

As Rheticus pursued his long journey from Varmia back to Wittenberg in the early autumn of 1541, he must have felt the magnitude of the task that lay before him: the delivery of a startling new theory of the whole physical universe.

Rheticus would have to perform this task with fewer social resources, and much less free time, than he had enjoyed in Varmia. Across Prussia there had been an established network of people for whom Copernicus was a man of stature as a person, as a doctor, and as a canon—all independent of his authority as an astronomer. That social credibility and connectedness had opened many doors for Rheticus, both personal and professional.

Schöner and Petreius were waiting expectantly in Nuremberg to hear the news from the north; and Rheticus's old friend and mentor Achilles Gasser, far off in the southwest, could be counted on for friendship and encouragement. But in the turbulent heartland of Protestant Germany, to which Rheticus was now returning, there was no one to compare with the mayor of Danzig or the duke of Prussia when it came to practical support. And there was no Copernicus. For all his optimism concerning the manuscript carefully packed into his luggage, Rheticus was about to face a peculiar kind of loneliness.

It was not only that people were going to have difficulty understanding him, but also that Rheticus himself could scarcely grasp the trials and stresses that his friends and colleagues had endured in his absence.

His desired coup on behalf of Copernican theory, beyond having *The Revolutions* published, involved winning the support of leading figures in Protestant Germany. Principal among these was his academic patron Melanchthon, the most influential German educator of the sixteenth century. In the previous year, even before the *First Account* emerged from the presses, Rheticus had arranged for his friend Andreas Aurifaber, then living in Danzig, to send the first printed sheets of the book to Melanchthon. The return of Rheticus had been preceded by the letter of support from the northern Protestant Albrecht of Prussia, with explicit greetings to the three most prominent men in Wittenberg: Luther, Melanchthon, and Johannes Bugenhagen. For Rheticus, that endorsement offered some reasonable hope that the doors of influence were about to swing open.

In 1541, however, Rheticus scarcely realized how close Melanchthon was to the limits of his mental and physical endurance. A year earlier he had been maneuvered into condoning the bigamy of Philip, landgrave of Hesse—a scandal that had damaged both the morale and the credibility of the reformers. Then, through the spring of 1541, Melanchthon and his colleagues, with the encouragement of Emperor Charles V, were engaged in strenuous, dangerous, and potentially historic negotiations with representatives of the pope aimed at possible reconciliation with the Catholic Church.

These efforts culminated in the Diet of Regensburg (Ratisbon). There, in April of that year, Melanchthon received a letter from Rheticus's former fellow student Paul Eber, also a professor in Wittenberg, reporting that "Joachim writes from Prussia that he is awaiting the completion of his teacher's work and cannot return for the coming semester [i.e., summer 1541]. He will be back in the autumn."[1] Under the circumstances, it must have seemed to Melanchthon, exhausted and heartsick at the critical but fruitless negotiations, that Rheticus was fiddling while Wittenberg burned.

Even after Rheticus returned to the university in early October, the tyranny of the urgent continued to dominate those who in less troubled times might have given the Copernican message the attention it deserved. In the absence of decisive empirical evidence, and unsupported by any mathematical proof confirming the heliocentric theory, Rheticus's enthusiasm seemed, in the strictest sense, impertinent.

No one was about to accept a counterintuitive notion like that of a moving earth merely on the word of an idiosyncratic, long-truant twenty-seven-year-old, at a time when the whole Protestant movement was busy trying to live down accusations that it encouraged gratuitous novelty and irresponsible teachings. Melanchthon, in a letter written shortly after Rheticus's return, added to an exasperated list of absurdities plaguing churches and schools the assertions of "the Polish astronomer who moves the earth and immobilizes the sun."[2]

Rheticus's hopes for a sympathetic reception of Copernican ideas were dashed, and so was his dream of getting a seventh consecutive semester off so that he could carry on to Nuremberg and begin publication of *The Revolutions* without delay. Instead, faced with an overeager faculty member bent on pursuing unsettling ideas, Wittenberg put him to work in administration.

In October of 1541 Rheticus found himself honored by his colleagues and superiors with election to the post of dean of Wittenberg's faculty of arts. As part of this role, in addition to various kinds of administrative work, he was expected to uphold high standards for both professors and students, and to offer lectures within his discipline. Although Rheticus was a professor of "lower mathematics," he opened the semester in October by posting an advertisement for a series of lectures on astronomy—with an emphasis on Ptolemy.*

The slightly grandiose advertisement opened with one of Rheticus's favorite proverbs: "Of all things, peace is the best" (*Pax optima rerum*). It then went on to lament the falling away of learning since ancient times, to assert the necessity of "literature and the arts" for human life, and to call Ptolemy's *Almagest* "by far the most beautiful among works of human hands."[3] Rheticus concluded his advertisement by laying out a canvas whose scope was Copernican, even if its subject was Ptolemaic: "It is absurd to presume that God the Architect established these wondrous and precisely regulated manifold motions in vain. It is fitting therefore that we should cultivate the science of these motions . . . God gave the human race shadows to be schoolmistresses of these things; moreover, he gave us numbers and measures so that we might discern how great a mind has constructed this amazing machine, and so that we should seek and cherish him. Accordingly, I shall begin the interpretation of Ptolemy next Thursday."[4]

The exaggerated reverence for Ptolemy expressed in the advertisement was no indication that Rheticus was shunning Copernicanism. In keeping with the belief that discovery proceeds by means of rediscovery, both Rheticus and Copernicus paid lip service to Ptolemy, and they offered their own work as the astronomical equivalent of friendly amendments to the efforts of their forebears. In the early 1540s, especially during Rheticus's first semester back in Wittenberg, public politeness as well as a lingering degree of wishful thinking kept his public statements from openly challenging Ptolemy.

One of Rheticus's formal duties during his deanship, from October

---

*Astronomy was sometimes restricted to "higher mathematics" (almost synonymous with astronomy). The professorship in this discipline at Wittenberg was held by Erasmus Reinhold, who would go on to use Copernicus's work without actually adopting Copernicanism. See Robert S. Westman's seminal article "The Melanchthon Circle, Rheticus, and the Wittenberg Interpretation of the Copernican Theory," *Isis* 66.2 (June 1975): 164–193.

1541 to the end of April 1542, was to preside over graduations. Two of these, in February and April, included the granting of the degree of master of arts.* In the first of his commencement orations, titled "On Astronomy and Geography," Rheticus defined astronomy as "the science that teaches the laws of the movements of the heavens," and he exhorted graduates "to incite and rouse" their own future students to pursue this discipline, adding, "We should be reminded of the Architect when we contemplate heaven."[5] As in his earlier writings for the duke of Prussia, Rheticus emphasized the interconnectedness of astronomy and geography. As befitted its formal occasion, the oration offered Rheticus a stage upon which he could display his religious and mathematical loyalties and urge a new generation of academics to recognize astronomy's contribution to both piety and civilization.

Rheticus's other graduation speech, in April 1542, accompanied the awarding of only ten new M.A. degrees, compared with twenty-two in February. However, the recipients included the illustrious Johannes Crato, who would prove himself a loyal friend to Rheticus, and who went on to become the personal physician to three successive emperors.

At this graduation Rheticus's topic was physics. Like the oration on astronomy and geography, this one delivered moral justification for learning generally—and again for learning as a foundation for civilization. "The human mind is an understanding nature, and indeed in being born it brings with it some unchangeable distinction between what is worthy and what is vile." Rheticus concluded by emphasizing an obligation to which he would return again and again throughout his career: "Let us rouse ourselves . . . to preserving the beneficial teaching for posterity."

Throughout his duties as a lecturer and administrator, Rheticus never forgot his own obligations as executor of the bequest of Copernicus.

---

*In the sixteenth century the M.A. was a professional distinction licensing its recipients to teach at the university level.

ORATIO-
NES DVAE, PRIMA
de Aſtronomia & Geographia,
Altera de Phyſica, habitæ
Vuittebergæ à Ioachimo
Rhetico, profeſſore
Mathematũm.

Norimbergæ apud Ioan. Petreium.

Title page of published graduation speeches given by Rheticus as dean
of arts at the University of Wittenberg.

He remained committed to delivering *The Revolutions* for publication
and kept up correspondence with his principal Nuremberg contacts:
Schöner and the printer Petreius.

During the late winter or early spring of 1542, Rheticus also found
time to cultivate the passion that would become his trademark: triangles.
Bypassing wider issues of Copernican cosmology and the theory of the
moving earth, he excerpted the trigonometric part of *The Revolutions*
(chapters 13 and 14) and had it published, in Wittenberg, under the

title *On the Sides and Angles of Triangles (De lateribus et angulis trian-gulorum)*.

The most interesting part of this publication, at least for nonmathematicians, is the dedicatory preface, addressed to Georg Hartmann, whom Rheticus had met in Nuremberg three and a half years earlier, in the autumn of 1538. Hartmann was among the most significant scientific figures of the day and, like Schöner, had encouraged Rheticus to visit Copernicus in the first place. He had known Copernicus's elder brother, Andreas, in Rome, and he would certainly be a prize convert to Copernicanism if Rheticus could win him over.

Rheticus opened his dedication of *On the Sides and Angles of Triangles* with the usual reflections about the lamentable falling away of learning since ancient times and the beneficial persistence of the liberal arts. He then went on to sing the praises of geometry as a means for hearing "the harmony of the celestial motions," while declaring the usefulness of the science of triangles "in all things pertaining to geometry, but especially in astronomy." Deftly introducing "the most illustrious and highly learned Mr. Nicolaus Copernicus" as an explicator of Ptolemy, Rheticus emphasized that his teacher "wrote most instructively concerning triangles."

Reaching its climax, the preface offered one of the most heartfelt tributes to Copernicus that Rheticus ever penned. "Such a learned man as you," he wrote to Hartmann, "will love this author equally for his brilliance and his learning, above all as regards the science of the heavens. In this he is comparable to the greatest creative minds of old." As for himself, Rheticus declared warmly, "There has been no greater human happiness than my relationship with so excellent a man and scholar as [Copernicus] is. And should my own work ever make any contribution to the general good (to the service of which all our efforts are directed), it shall be owing to him."[6]

In 1542, as the winter semester wore to a close, Rheticus prepared to resume his travels once more and to take another sabbatical leave from his academic post. Rumors were already in circulation about a possible professorship for him in Leipzig, but first he had to travel south to Nuremberg and to the printing house of Petreius, who was waiting to publish Copernicus's manuscript. Rheticus knew that setting and printing a work as complicated as *The Revolutions* would take time, so he planned to pursue other projects as well while in Nuremberg. He also hoped to venture beyond Nuremberg and visit friends and family in Feldkirch and the surrounding area.

Before he did that, however, Rheticus spent almost a month renewing acquaintances with his prominent Nuremberg contacts. He had already dedicated his two previous Copernican publications, the *First Account* and *On the Sides and Angles of Triangles*, to Schöner and Hartmann respectively. For Rheticus, only twenty-eight years old, revisiting these men brought a sense of pride and accomplishment. Hartmann showed his respect by entrusting Rheticus with the manuscripts he himself had inherited from Johannes Werner: "Spherical Triangles" and "Observations on the Heavens."

During the first half of June, Rheticus set off from Nuremberg and continued southwest to his home region of Rhaetia. He visited his mother, Thomasina, living in Bregenz, having married Georg Wilhelm, now that city's mayor. In his mother's house Rheticus also spent time in contact with his stepfather's son Bartholomeas, who later would enroll in the University of Leipzig. Rheticus also visited his sister Magdalena in Ravensburg, and his old friend and mentor Achilles Gasser in Feldkirch. True to his endearing habit of bearing books as gifts, Rheticus gave Gasser a copy of the recently published *On the Sides and Angles of Triangles*.* And true to another habit, he set about cultivating contacts with as many important people as time allowed.

---

*This volume is held today in the library of the Vatican. In the front of the book, Gasser carefully recorded the name of the giver and the date of the gift, June 20, 1542. Three months later, after clearing out some of his late father's remaining possessions from the former family house in Feldkirch, Rheticus

One of the people he met was Heinrich Widnauer, mayor of his hometown. This acquaintance would later afford Rheticus, after his return to Nuremberg in midsummer of 1542, an ideal opportunity to express a range of his deepest obligations—all in the process of some dignified showing off. In August of 1542, with *The Revolutions* making its way through the press, Rheticus had Petreius print the two graduation speeches he had delivered earlier that year in Wittenberg, and he dedicated the publication to Widnauer. Rheticus began the letter with effusive praise for the liberal arts in which, using the beloved metaphor of music, he set forth his conviction concerning the divine interrelatedness of all the arts and sciences: "The persistent harmony of the laws, whereby one precept as it were gives birth to the next, demonstrates clearly the truth of the claim that this harmony is traceable to its own sacred origin in the creating hand of God."

In this dedication Rheticus went on to praise his childhood teachers, including his father, and then acknowledged a man who was still playing a major role in his career (and who, he knew, would himself soon read these words): "my teacher Philipp Melanchthon, that singular treasure of our age." This learned man, "seeing I had already acquired a modest knowledge of numbers, and perhaps for other reasons unknown to me, recommended that I enter the field of mathematics," recalled Rheticus. It was at Melanchthon's urging, then, that Rheticus had taken up the pursuit of this discipline with such "great diligence."

Rheticus also related how he studied geometry and astronomy with the Wittenberg professor Johannes Volmar (originally from near Feldkirch), "our countryman," and how an interest in these disciplines subsequently led him to visit Schöner in Nuremberg and Joachim

---

gave Gasser still another book, one devoted to a medical discussion of "the French disease" (i.e., syphilis). See Karl Heinz Burmeister, "Georg Joachim Rhetikus—ein Bregenzer?," *Montfort: Vierteljahresschrift für Geschichte und Gegenwart Vorarlbergs* 57.4 (2005): 311.

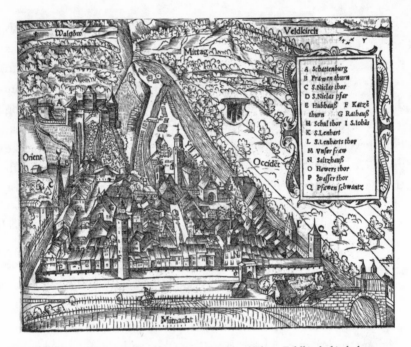

*Sebastian Münster's depiction of a triangular-looking Feldkirch, birthplace of Rheticus (*Cosmography, *1550).*

Camerarius in Tübingen. These points all served, however, as prelude to the dedication's Copernican climax.

> Finally, hearing the great fame of Dr. Nicolaus Copernicus in the far north, even though the University of Wittenberg had appointed me professor in those disciplines, I knew I should have no rest until I my-self learned something of his teaching. And indeed I regret neither the expense, nor the long journey, nor any of the other hardship. Rather, I feel I have reaped a great reward. For by means of a certain youthful audacity I was able to spur this eminent man on to communicate to the whole world his theories regarding that subject earlier than might have been. And all learned minds will join in my assessment of these theories as soon as the books we now have in press in Nuremberg are published.[7]

Rippling beneath the surface of the letter to Widnauer are some of the competing tensions that continued to mold the psyche of Rheticus. If the letter were a play, three separate father figures would appear on stage (a fourth, the mayor himself, sitting in the audience), while the hero, Rheticus, strives to please or honor all of them at once. Of course, the real father, the now-nameless and long-dead Iserin, is beyond being pleased or displeased, and to honor him is enough. As for Copernicus, the most recently and fervently adopted father figure, there is also no doubt: Rheticus honors him and is confident he can make him proud.

But the third corner of this paternal triangle, Melanchthon, was poised between continued support and admiration for Rheticus on the one hand, and impatience and censure on the other. And so, in his published letter to Widnauer, Rheticus took care to remind Melanchthon that he held him in high honor—and also that he, Melanchthon, shared some responsibility for Rheticus's devotion to mathematics in the first place. Rheticus's uncertainty regarding Melanchthon's continued paternal support made him all the more zealous to stress the value of the goals he was pursuing. Copernicus had done the same thing, a generation earlier, when he sought to avert the disappointment of his own main father figure, his uncle Lucas Watzenrode, not wanting to prove deficient in his eyes.

It is hard to say exactly what was creating the strain in Rheticus's relationship with Melanchthon. Despite the latter's grumpy comment the previous autumn about "the Polish astronomer who moves the earth and immobilizes the sun," Melanchthon continued to support Rheticus warmly and generously. Upon Rheticus's departure from Wittenberg in May of 1542, Melanchthon endorsed the traveler's reentry into the highest circles in Nuremberg with a letter to one of that city's most influential theologians, Veit Dietrich, recommending Rheticus to him "and other friends" as a man "well suited to teaching those sweetest arts concerning the motions of the heavens."[8]

About the same time, there was growing talk about Rheticus's possible relocation to the University of Leipzig. Joachim Camerarius, who had himself recently moved there from Tübingen, was intent on building up Leipzig's academic strength and on luring the young mathematics professor away from Wittenberg. Yet Melanchthon unselfishly wrote Rheticus a letter of reference for this undertaking too.

Melanchthon also continued to play a personal role in support of Rheticus's scholarly research. In June of 1542 Rheticus—knowing his mentor's humanistic interests—wrote Melanchthon to tell him of a Greek manuscript on conic sections written by the third-century-BC mathematician Apollonius of Perga. Melanchthon responded at once by writing a Nuremberg senator, Erasmus Ebner, requesting that the authorities release this manuscript from the collection left behind by Regiomontanus so that Rheticus might prepare a scholarly edition of it.[9]

Despite Melanchthon's generous academic support—even mentorship—something about Rheticus was causing the elder man to feel uneasy about this eager young professor with the wandering ways. Hints about that uneasiness appear in a single letter that Melanchthon sent in the summer of 1542, on July 25, about the time that Rheticus was returning to Nuremberg from his family visit to Feldkirch and environs.

Melanchthon, writing again to Camerarius in Leipzig, commented first on a characteristic displayed by Rheticus, perhaps related to his youth, that he had had to put up with: a certain excitability and "enthusiasm" for one kind of (unnamed) natural philosophy.* Melanchthon added that he would prefer Rheticus took a more balanced, "Socratic" approach—that is to say, he wished he would find a wife and become a "paterfamilias."[10]

---

*Until recent times the word *enthusiasm* had quite negative connotations, rather like *obsession* today. It bespoke imbalance and lack of rational control.

Melanchthon knew whereof he spoke. He himself had in his twenties been obsessed with his own academic work and had therefore not wanted to marry. But his superior, Martin Luther, had persuaded him to take a different path, and he had gradually come to see his marriage as one of the wisest decisions of his life. He had left Rheticus in no doubt that he too should follow the reformers' well-established example. And Rheticus, always keen to please his *praeceptor*, took the lesson to heart. Writing in June of 1542—in a conscious display of balance— Rheticus told Melanchthon not only of having found the manuscript that he wished to edit, but also of hoping to find a woman whom he could wed.

Thus the second piece of news that Melanchthon communicated to Camerarius in his letter of July 25 concerned Rheticus's professed intention to marry. The benevolent Melanchthon had permitted himself to betray his dissatisfaction with Rheticus only because he believed that the condition would soon be remedied. Considering Rheticus's own interests and well-being, Melanchthon could not imagine a more apt prescription for a happy, balanced life than marriage.

But he did not know Rheticus as well as he thought he did, and perhaps neither did Rheticus—though it was not out of character for him to agree to look for a woman out of a desire to please a man, especially a father figure like Melanchthon. It is also possible that the ambitions and expectations of Rheticus's mother, Thomasina, played a role.

Although details of the story are obscure, it is known that in 1542 Rheticus made a second trip back to the Feldkirch area. It is likely he was turning once more to old friends to help him decide whether he should accept the position offered at Leipzig. In an undated letter from this period Melanchthon mentioned that he had written to Rheticus urging him to make up his mind and to communicate his decision clearly.[11] On September 18, 1542, there is record of Rheticus visiting with his old mentor and friend Achilles Gasser in a house owned by his mother, which Thomasina had purchased after remarrying and moving from Feldkirch to Bregenz.*

Apart from Melanchthon's letters and other fragmentary documentation of this period, there is one further piece of intriguing circumstantial evidence suggesting that Rheticus may have returned home to get married. In the city archive of Lindau, not far to the northwest of Bregenz along the shore of Lake Constance, can be found the parish baptismal records for the years 1534–67. In this volume appears an entry, dated October 18, 1542, marking the baptism of the first daughter of an old friend of Rheticus's from Wittenberg, Johannes Heldelin, a Lindau schoolteacher. The entry includes the names of two godparents: "Jerg Jochim von Veldkirch"—Rheticus, who at the time was twenty-eight—and Candina Mürgel, daughter of a Lindau doctor, Johannes Mürgel, who is thought by some to have been the original owner of the copy of Ptolemy's *Almagest* that Rheticus carried with him to Frauenburg as a gift for Copernicus.[12]

Their common social circle in Rhaetia, their proximity and participation as godparents at an event as religious and intimate as a baptism, their shared confession (Lutheran), the fact that Rheticus was still lingering in Lindau when his academic duties were about to begin elsewhere—all of these details render plausible the suspicion that Candina was Rheticus's intended bride. Add to this the remarkable fact that there is—apart from his mother, his sister, and a few colleagues' wives—only one woman whose name ever appears anywhere in the extensive record of Rheticus's life, and that is Candina Mürgel.

In the end, Rheticus did not marry Candina, or whoever else his intended might have been. It is unclear why he did not, and unclear whether even Rheticus himself knew, in any conscious way, the nature of his deepest sexual motivations.

---

*The house in Bregenz, at Kirchstrasse no. 13, still stands today. See Karl Heinz Burmeister, "Neue Forschung über Georg Joachim Rhetikus," *Jahrbuch des Vorarlberger Landesmuseumsvereins*, 1974–75 (Bregenz: Freunde der Landeskunde, 1977), p. 41.

CHAPTER 8

WELCOME TO LEIPZIG

In the autumn of 1542, the world outside was making urgent demands that Rheticus could no longer evade. Not only was he unable to face marriage; he also, it seems, could no longer face Melanchthon and Wittenberg. His preference would have been to avoid academia altogether for a while longer and return to Nuremberg to see *The Revolutions* through the press. That, however, would mean losing his chance for a position at Leipzig and having to ask Wittenberg for a further leave of absence—a dim prospect given the darkening mood of Melanchthon.

Indeed, Melanchthon wrote another letter (this time partly in Greek, to keep it confidential) warning Camerarius that, should Rheticus accept the Leipzig professorship, unambiguous terms should be spelled out regarding both his salary and his duties. Melanchthon's intention was to be honest and helpful, not mean. In further correspondence with Camerarius (November 18), after Rheticus had taken up his new position, Melanchthon's kindness of intent is evident, despite a continued undertone of caution: "I commend Joachim Rheticus to you. Do support him with words of most faithful counsel."[1]

The matriculation record for winter semester 1542 shows that

"Joachim Rheticus, MA, was made professor of mathematics."*[2] The fact that another colleague mentioned in the same entry, Balthasar Klein, was made professor of "lower" or "rudimentary" mathematics demonstrates that Rheticus's new position was no lateral move. In Wittenberg he had held the chair of lower mathematics, while Erasmus Reinhold was professor of higher mathematics (i.e., astronomy). In Leipzig Rheticus was now for the first time a fully professional astronomer in title as well as in fact.

If Rheticus himself had somewhat mixed feelings at his installation in this new position, the same cannot be said of the University of Leipzig. Parallel to a pattern familiar in academia today, Leipzig was an institution determined to move up in the world, and therefore trying to attract faculty who were the brightest and the best. Accordingly, Rheticus was recruited not only because he was outstanding in his field, but also because Leipzig apparently craved the satisfaction of taking him away from a competitor institution that was reluctant to lose him.[3] From the Leipzig faculty's point of view, the happy result was that it "lured away from Wittenberg the eminent and most learned gentleman professor Joachim Rheticus."[4]

To accomplish this, Leipzig was ready to pay—and Rheticus was ready to be paid. The university records of 1542 state that the going rate for a professor in the faculty of arts was about a hundred florins per annum, probably what Rheticus was paid at Wittenberg. The records likewise state, however, that "when he would not and could not be content with this amount," the offer was sweetened to 140 florins. In keeping with the oddly circular logic to which institutions are prone,

---

*On the way to Leipzig to begin his new job, Rheticus was joined by an old friend, Caspar Brusch. Upon arriving in Leipzig, the two young men took lodging together at the Inn of the Golden Cross. In the next century, during the Thirty Years War, this inn was renamed The Fireball (*Zur Feuerkugel*); much later still, from 1765 until 1768, one of its tenants would be a budding young poet by the name of Johann Wolfgang von Goethe.

Leipzig congratulated itself on having hired a professor worth 40 percent more than those teaching at other universities. Amid considerable celebration, Rheticus was presented by Camerarius to the entire senate of the faculty of arts and duly received "as professor of higher mathematics by unanimous consent."

Rheticus was by any standards a model of academic success: only twenty-eight, and paid an excellent salary by a prestigious university to teach at the highest level of his discipline. Yet he was exhausted. While residing in Varmia, Rheticus had devoted himself to the intense labor and excitement of absorbing Copernicus's teaching—in addition to publishing the *First Account*—along with local travel to Löbau, Danzig, Königsberg, and other Prussian locations required by his cartographic research. Since then, in addition to academic duties, he had made multiple journeys to Nuremberg and to his home region in the southwest, before making his way to Leipzig. For four years he had been almost constantly on the move.

Apart from his physical exhaustion, however, Rheticus's temperament never allowed him to take much pleasure in institutions. Although he was highly sociable, loved people, cherished humanistic learning and academic conversation, and could be a loyal friend, he was by nature a maverick. Repeatedly throughout his life, he showed that he could not, or would not, establish any long-term commitment to a school or university. Institutions had their uses, but none could capture his mind or his heart.

During the autumn and winter of 1542–43, what captivated Rheticus was not his new position but a book whose pages were being typeset and collated in the printing house of Johannes Petreius in Nuremberg, two hundred miles south of Leipzig. The work that his beloved teacher had entrusted into his hands was now, awkwardly, in the hands of others, and all that Rheticus could do was keep himself busy with his new academic duties and hope that everything in Nuremberg would turn out well.

Whether things in Nuremberg did turn out well grew into one of the most hotly debated questions in the history of science. The issue centered on the Nuremberg theologian and avid amateur mathematician Andreas Osiander, whom Rheticus trusted to oversee the final stages of publication of *The Revolutions*. After becoming a priest in the early 1520s, Osiander converted to Lutheranism and became Nuremberg's first Protestant minister. This thoroughly German man cultivated a considerable international network that included the soon-to-be Archbishop of Canterbury, Thomas Cranmer.* Theologically, Osiander had a reputation for controversy and stubbornness. But what actually earned him the enmity of Rheticus were his efforts to be overtly cosmologically accommodating and diplomatic.

Rheticus had probably first met Osiander back in 1538 upon entering the Nuremberg circle of Schöner and Petreius. Osiander's own considerable network was useful to these men, for it heightened the international visibility of Nuremberg, and with it that of Petreius's publishing business. Accordingly, like Petreius himself, Osiander viewed Rheticus's expedition to Varmia in 1539 as promising closer links with a fresh source of learning—and possibly also of profit and prestige.

The fragments that survive from Osiander's 1540 correspondence with Rheticus make the high-ranking, middle-aged theologian appear surprisingly deferential toward the young mathematician. "I ask you over and over again," Osiander had written imploringly to Rheticus in Frauenburg, "just as you offer me your friendship, in the same way to exert your efforts so as to obtain the friendship of this man [Copernicus] for me too." Rheticus obliged by sending several copies of his *First Account*—presumably for distribution among the learned and powerful of Nuremberg. Osiander responded: "[The *Account*] pleased me very

---

*Cranmer had visited Osiander in 1532 and married his niece. In the 1540s Osiander worked closely with the Italian astrologer-mathematician Girolamo Cardano and in 1549 moved north to Königsberg to take up the chair of theology at the new university founded by his and Rheticus's mutual supporter Duke Albrecht of Prussia.

much. The book has very clear introductory discussions of topics to be expected hereafter."[5] The two men's relations seemed cordial enough. But the seeds of trouble are visible in two letters that Osiander wrote the following year, both dated April 20, 1541, and sent to Frauenburg: one of them addressed to Copernicus, the other to Rheticus.

In the letter to Copernicus, Osiander, for all his professed interest in science, showed that he was neither a scientist nor a true natural philosopher. He instead belonged to the old school of astronomers who knew that their devices did not correspond with physical reality, but who cultivated these devices anyway because they saved the appearances—that is to say, provided a mechanism for calculating observable stellar and planetary positions. He therefore urged Copernicus to offer his work merely as a useful hypothesis: "I have always felt about hypotheses that they are not articles of faith but the basis of computation. Thus, even if they are false, it does not matter, provided that they reproduce exactly the phenomena of the motions . . . It would therefore appear to be desirable for you to touch upon this matter somewhat in an introduction. For in this way you would mollify the peripatetics and theologians, whose opposition you fear."[6]

Because Copernicus considered his proposal that the earth revolves around the sun to be more than just a hypothesis, he had been anxious about the reactions of two sorts of people: theologians (since some passages of Scripture, read literally, seemed to imply that the sun moves and the earth stands still); and "peripatetics," those Aristotelian philosophers who dominated the study of physics in the universities. Osiander, trying to be helpful, suggested a way of avoiding the difficulty and mollifying these critics. As he wrote to Rheticus, a preface offering Copernicus's work merely as a hypothesis—rather than a physical proposal—could in effect buy time for heliocentrism. Potential opponents would then "be diverted from stern defense and attracted by the charm of the inquiry; first their antagonism will disappear, then they will seek the truth in vain by their own devices, and go over to the opinion of the author."[7]

Rheticus took seriously both sorts of opposition that Osiander had worried about, although objections based on passages of Scripture were—at least at that time—easier to dispel than were the philosophical kind. Before he left Varmia in 1541 Rheticus had composed his own small tract to demonstrate the absence of conflict between heliocentrism and the Bible.[8] He began the work by proposing, on the authority of St. Augustine, that any scientific obscurity should be tackled "by means of enquiry, not assertion" (*non affirmando, sed quaerendo*). He went on to make a distinction that is still part of the faith-science dialogue: In the Bible the Holy Spirit's intention, declared Rheticus, is not to teach science but to impart spiritual truths "necessary for our salvation." Moreover, whatever descriptions of nature do appear in the Bible are "accommodated to the popular understanding."[*]

Rheticus never published his work on the compatibility of the Bible and Copernicanism. It appeared anonymously and inconspicuously in the Netherlands more than a century after its composition, and it was not discovered and identified as Rheticus's work until the early 1980s.[9] Copernicus's best friend, Tiedemann Giese, wanted it to be included in an improved, second edition of *The Revolutions*[10]—but it was not. In any case, in his preface to the pope, what Copernicus offered to possible opponents who raised Bible-based objections was not a counterargument but the back of his hand. He called them ignorant, "idle

---

*John Calvin, employing the same distinction, would endorse astronomers' proofs that (for example) Saturn is larger than the moon—even though Genesis identifies the sun and the moon as being the "two great luminaries." Similarly, early in the seventeenth century Kepler would defend heliocentrism by distinguishing between scientific and popular discourse. The Holy Scriptures "speak with humans in the human manner." So there is no contradiction. Even today people who at a scientific level reject geocentrism talk about the sun rising and setting. They are simply describing in everyday language what their eyes actually see. Both Calvin's and Kepler's comments are excerpted in *The Book of the Cosmos: Imagining the Universe from Heraclitus to Hawking*, edited by Dennis Danielson (Cambridge, MA: Perseus, 2000), chap. 20.

talkers" who distort and twist Scripture. He had "only contempt for their audacity."*

In the 1540s, however, the principal opposition to Copernicus would come not from theologians but from the philosophers who held sway in the universities and who continued to adhere to the teachings of Aristotle and Ptolemy. It was mainly their objections that Osiander was trying to circumvent when—entirely on his own authority—he inserted the letter "To the Reader Concerning the Hypotheses of This Work" ("Ad lectorem") as an anonymous foreword to Copernicus's *Revolutions*. He was concerned about possible offense not chiefly among clergy or common folk but among scholars loyal to "the liberal arts"—to the classically based curriculum of the schools and universities. Accordingly, he hoped to mitigate their concerns by suggesting that the hypotheses of Copernicus's work, "devised by the imagination, . . . are not put forward to convince anyone that they are true, but merely to provide a reliable basis for computation."[11]

When the first eagerly awaited copies of the great work arrived in Leipzig in the spring of 1543, Rheticus saw the "Ad lectorem" on its opening pages and he was devastated—then livid. He regretted that he had not stayed on in Nuremberg personally to shepherd *The Revolutions* through the press, and was chagrined that Petreius could have allowed such a betrayal of Copernicus's intentions. The problem was not with the word *hypothesis*. Rheticus himself had referred to "the hypotheses regarding the motion of the earth"—but had at once declared them to be so robustly "consonant with the appearances" that reason itself could postulate nothing "nearer the truth."[12] Copernicus was offering a new real world, yet here was this cowardly, equivocating "Ad lectorem" presenting it as just another useful but imagined set of fictional devices.

---

*It was only decades later, after the rise of the Counter-Reformation and the Council of Trent, when literalistic interpretation of the Bible had become the official Roman Catholic norm, that the apparent collision of heliocentrism and Scripture truly came to a head—as Galileo discovered.

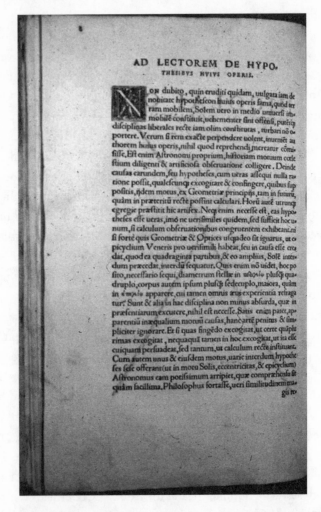

Andreas Osiander's anonymous preface to The Revolutions *(1543), crossed out by Rheticus. The right-hand page includes, undefaced, Cardinal Nicolaus Schönberg's 1536 letter requesting that Copernicus divulge his cosmology.*

gis requiret, neuter tamen quicquam certi compræhedet, aut
tradet, nisi diuinitus illi reuelatum fuerit. Sinamus igitur &
has nouas hypotheses, inter ueteres, nihilo uerisimiliores inno
tescere, præsertim cum admirabiles simul, & faciles sint, ingen
temq; thesaurum, doctissimarum obseruationum secum ad
uehant. Neq; quisquam, quod ad hypotheses attinet, quicquam
certi ab Astronomia expectet, cum ipsa nihil tale præstare que
at, ne si in alium usum conficta pro ueris arripiat, stultior ab
hac disciplina discedat, quàm accesserit. Vale.

NICOLAVS SCHONBERGIVS CAR
dinalis Capuanus, Nicolao Copernico, S.

Vm mihi de uirtute tua, costanti omniu sermone
ante annos aliquot allatu esset, cœpi tum maiorem
in modu te animo coplecti, atq; gratulari etia no
stris hominibus, apud q̄s tata gloria floreres. Intellexera enim
te nō modo ueteru Mathematicoru inueta egregie callere, sed
etiā nouā Mūdi ratione costituisse. Qua doceas terrā moueri:
Solem imu mūdi, adeoq; mediu locu obtinere: Cœlu octauu
immotu, atq; fixu ppetuo manere: Lunā se una cu inclusis suæ
sphæræ elementis, inter Martis & Veneris cœlu sitam, anni
uersario cursu circu Solem couertere. Atq; de hac tota Astro
nomiæ ratione comentarios à te cofectos esse, ac erraticarum
stellaru motus calculis subductos in tabulas te cotulisse, maxi
ma omniu cum admiratione. Quamobrem uir doctissime, ni
si tibi molestus sum, te etia atq; etia oro uehementer, ut hoc
tuu inuentu studiosis comunices, & tuas de mundi sphæra lu
cubrationes unā cu Tabulis, & si quid habes præterea, q̄d ad
eandem rem pertineat, primo quoq; tempore ad me mittas.
Dedi autem negotiu Theodorico à Reden, ut istic meis sum
ptibus omnia describantur, atq; ad me transferantur. Quod si
mihi morem in hac re gesseris, intelliges te cum homine no
minis tui studioso, & tantæ uirtuti satisfacere cupiente rem ha
buisse. Vale, Romæ, Calend. Nouembris, anno M.D.XXXVI.

ij

Rheticus, in a deliberate act of protest and defiance, took his beloved teacher's newly arrived work and opened the precious book to the two facing pages where the "Ad lectorem" appeared. Seizing a red crayon in his right hand, he carefully defaced Osiander's letter with a pair of large, angry freehand *X*'s. He did the same with each copy that came into his possession, including the ones he sent as gifts. Everyone who saw the book could sense, too, how Rheticus's great pride at the appearance of Copernicus's work was tainted with deep disappointment and frustration.

As for Copernicus himself, although he did view a printed copy of *The Revolutions*, it was only at his last breath on his dying day. Tiedemann Giese, however, whose personal correspondence with Rheticus recounted the circumstances of the death of Copernicus, immediately noticed the "Ad lectorem" and was every bit as outraged as Rheticus.

Giese fought back not with the wordless censure of a crayon but with the acid eloquence of his pen. As he wrote to Rheticus: "At the very threshold [of the book] I perceived the bad faith and . . . the impiety of Petreius, which produced in me an indignation more bitter than my previous sorrow. For who will not be anguished by such a scandal committed under the cover of good faith?"[13] Having alleged that the "Ad lectorem" was virtually a case of sacrilege, though, Giese went on to doubt whether it was Petreius who ought to be blamed for the "deception," or whether he was merely negligent in letting someone else "diminish faith in the treatise." Regardless, Giese assured Rheticus that he was writing a formal statement of complaint to the city council of Nuremberg. It was in this context that he began to contemplate a new edition of *The Revolutions* with prefatory material by Rheticus that would cleanse the work of "the stain of chicanery."

Giese's notice of complaint was duly forwarded to the Nuremberg council, and Petreius fought back like any man stung with a false accusation. In a note that has been preserved, the council sent its secretary, Jerome Baumgartner, the following instruction regarding Petreius's response and the council's decision not to take action against him: "Send to Tiedemann [Giese] . . . the answer written by Johannes Petreius to the bishop's communication (in the answer, the acerbities should be omitted and mitigated). Add to it: no punishment can be inflicted on Petreius in this matter on the basis of his answer."[14]

Meanwhile, Osiander kept above of the fray. Giese as of yet had little knowledge of him, and Rheticus only suspected his involvement. But Osiander's authorship of the offending piece would be attested years later through a chain of communication leading from Osiander himself to Peter Apian and then through his son Philip Apian (1531–89) indirectly to Michael Maestlin (1550–1631) and Johannes Kepler. To Peter Apian, Osiander openly admitted, late in 1548, that he had conceived and written the "Ad lectorem." Maestlin, finding reference to the matter among Philip Apian's papers, transcribed the following account into his own copy of *The Revolutions*.

> Concerning this letter ["Ad lectorem"], I (Mich. Maestlin) found the following words written somewhere among Philip Apian's books (which I bought from his widow) . . . "On account of this letter, Georg Joachim Rheticus, the Leipzig professor and disciple of Copernicus, became embroiled in a very bitter wrangle with the printer. The latter asserted that it had been submitted to him with the rest of the treatise. Rheticus, however, suspected that Osiander had prefaced it to the work, and declared that, if he knew this for certain, he would sort the fellow out in such a way that he would mind his own business and never again dare to slander astronomers. Nevertheless, [Peter] Apian told me that Osiander had openly admitted to him that he had added this [letter] as his own idea."[15]

These words as recorded by Maestlin (who became Kepler's teacher) indicate the level of Rheticus's fury. It is the only existing evidence that Rheticus ever came close to meditating or threatening violent acts against another person. Yet there is no evidence that his sense of outrage was widely shared. Apart from Giese, Rheticus alone responded as if the "Ad lectorem" dishonored a sacred patrimony. Indeed, his friend Andreas Aurifaber, one of the recipients of a copy of *The Revolutions* with Osiander's preface personally crossed out by Rheticus, a few years later married Osiander's daughter.[16] Giese was dead wrong in assuming that everybody would be anguished by such a scandal.

Few things reveal character as clearly as how a person deals with disappointment. And to his credit, there is no evidence beyond the summer of 1543 that Rheticus wasted any time nurturing his grievance over the "Ad lectorem." Without quite admitting it, perhaps he realized that Osiander was partly right: He *was* buying time for Copernicus. For those lacking the equivalent of a conversion experience in the presence of the master himself,[17] there were just too many blanks to be filled in before they could accept his theory. As one observant scholar put it, "Copernicus's position clearly requires a new physics but does not provide it."[18] Rheticus already knew that extensive further observational support was required, and also that sound interpretation of those astronomical data demanded more powerful mathematics than any yet available. He himself had seen the wisdom of offering Copernicus's trigonometry even before the publication of *The Revolutions*.

These advances would take time and patience. Yet Rheticus was still only twenty-nine. He enjoyed good health and (judging from all the traveling) a robust constitution. He had a fine academic reputation, and he had important academic duties to perform at Leipzig. These obligations seem to have kept him completely occupied for two years beyond the publication of *The Revolutions*, two years beyond the death of Copernicus.

Amid the academic activity, however, punctuated by private moments of grief and reflection, Rheticus devoted some time to planning his next journey; before the end of summer semester 1545, he set off on the road to Italy.

CHAPTER 9

GAMBLING ON CARDANO

In his letter to the pope at the beginning of *The Revolutions*, Copernicus had declared, a little defensively, that "mathematics is written for mathematicians."* But with Copernicus gone, skilled help with mathematics was exactly what Rheticus needed. It was quite logical, then, that he should seek it from a mathematician whose star was rising like none other.

Girolamo Cardano (1501–76), thirteen years older than Rheticus, was one of the most weirdly colorful and internationally famous scholars of the sixteenth century. Born in Pavia, near Milan, after failed attempts to abort him, he survived to become a doctor, astrologer, mathematician, and (influenced by his gambling habit) founder of the science of probability.[1] His father, Fazio, a lawyer and mathematician, was a friend of Leonardo da Vinci. Cardano's illegitimate birth was used as an excuse to bar him from the envious Milanese College of Physicians during the 1520s and 1530s. However, his fame as a doctor extended to the farthest reaches of Europe, and in 1552 he was invited to travel all the way to Scotland, where he successfully treated the asthmatic archbishop of St. Andrews.

Mathemata mathematicis scribuntur. Because of the close association at the time between mathematics and astronomy, this statement is sometimes translated "Astronomy is written for astronomers."

The international scope of Cardano's fame was reinforced by a professional relationship he established with the printer Petreius in Nuremberg. It was this link, especially as it related to astrology, that initiated the acquaintance between Cardano and Rheticus.* Some of Rheticus's own publications straddled popular and professional astrology, and Cardano first mentioned his "friend Georg Joachim"—in a work published by Petreius in 1543—as someone who had provided him with horoscopes.[2]

This professional and literary thread, together with the Italian roots of Rheticus's own family, may have made a visit to Cardano in Milan in 1545 seem appealing.[3] What clinched Rheticus's decision to head south, however, was Cardano's growing reputation as a mathematician. Not only did such a visit promise to bring Rheticus abreast of the most leading-edge mathematical thought of the day; it also held out the prospect of possibly winning over to Copernicanism the world's leading mathematician. The authority of such a supporter could compensate for the evidential deficit that Rheticus knew was still holding back the progress of Copernicanism.

That crucial deficit was illustrated in the early 1540s by two quite different readers of Rheticus's work. The first of these was the famous Dutch cosmographer Rheticus referred to as "the most learned Gemma Frisius."[4] Gemma Frisius had received and devoured a copy of the *First Account* soon after its publication in 1540, and his response is recorded in a letter he sent to his patron the Varmian bishop John Dantiscus on July 20, 1541, before Rheticus himself headed back from Frauenburg to Wittenberg later the same year.

Writing in the florid epistolary style of the times, Gemma exclaimed that "Urania [the muse of astronomy] seems to have established a new

---

* Much later, in 1570, the Inquisition would throw Cardano in jail for having cast a horoscope of Jesus Christ—though presumably his 1562 work *In Praise of Nero (Encomium Neronis)* had not endeared him to the Church either.

residence there [in Varmia] and raised up new worshippers who are about to offer us a new earth, a new sun, new stars, indeed a whole new world." Gemma's enthusiasm was inspired by the prospect of doing away with the litany of deficiencies displayed by standard Ptolemaic astronomy: "[These include] errors, evasions, labyrinths—even greater enigmas than those of the Sphinx. I could certainly enumerate many that I find quite unsatisfying, such as the motion of Mars . . . And so, if that author of yours [Copernicus] should achieve a restoration [of this science] to a state of soundness and good repair (which my mind has been greatly anticipating since I received the advance account that was sent), then would not that indeed offer up a new earth, new heavens, and a new universe?"

Nevertheless, Gemma was not about to embrace heliocentrism as a cosmology. Instead, he was interested in Copernicus's astronomy for its results, not its truth. "I do not argue about the hypotheses which that [astronomer] uses in his account, whatever they be, or however much truth they may possess. Nor does it concern me whether the earth is said to revolve, or whether it stands still." What did excite Gemma was the promise of a system characterized by "very precise calculation" (*in exactissimum calculum redacta*). The prospect of such advances, he wrote, "[fills me with] desire to see this business brought to fruition. And everywhere there are more than a few erudite men whose minds desire it no less than I do."[5]

Gemma thus suggested the existence of an attentive audience for Copernicus's ideas, at the same time indicating how hard that audience would be to convince. For Rheticus, this response was at once challenging and disappointing—not so much because Gemma used the word *hypothesis*, as because he implied that heliocentrism might be merely a hypothesis, that its truth might not matter. For Rheticus, truth mattered. As defender of that truth, he faced the huge task of coming up with better, more accurate evidence; this meant assembling both the observations and the mathematical tools that would enable the "very precise calculation" that Gemma and others like him desired.

Achilles Gasser, another reader of Rheticus's work, also recognized the need for better mathematics. Rheticus had visited him in Feldkirch on his way south to Italy in 1545. Not long after those days of renewed friendship, Gasser published an astrological *Prognosticum*—a sort of almanac—for the coming year, 1546. This small work, published by Petreius, appeared in simultaneous German and Latin versions, each with a separate preface.[6]

The German edition, titled *Practica* and addressed to a Tyrolean nobleman named Caspar Täntzl, contained the first explicit mention of Copernicus in any publication in the German language. Gasser's preface informed Täntzl that "the most learned and wonderful man Dr. Nicolaus Copernicus, away off in Prussia, [teaches that] the sun . . . stands unmoved in the midst of the whole universe" and that "this earthly realm . . . variously courses round between the planets Venus and Mars." But Gasser also had the audacity to inform Täntzl that Copernicus had "demonstratively proven his theory among the mathematicians."

This was a wild exaggeration, and Gasser knew it. He personally seems to have accepted heliocentrism not because he fully grasped its soundness but on account of his own implicit trust in Rheticus. Yet he wanted more hard evidence than Rheticus had so far offered. Gasser sang two contrary tunes: in the dedication to Täntzl (in the German version) claiming that Copernicanism was already mathematically proven; but in the other dedication, written to Rheticus himself on the same day, begging to be provided with better reasons for accepting the new cosmology: "For that unremitting passion for the restoration of astronomy . . . implores you [Rheticus], with devotion constant and overflowing, to proclaim with no meager praise that new and paradoxical account of the celestial science, which, since the demise of that most illustrious gentleman Nicolaus Copernicus, your teacher, it is now up to you alone to do. Therefore, . . . offer an easier introduction to this science, as well as clearer proofs."[7]

Despite Gasser's inconsistency with himself, his call for "clearer proofs" was entirely consistent with Gemma Frisius's desire for "very precise calculation." From both admired scholar and personal friend, the message was unmistakable: If Copernicanism was to be accepted, stronger mathematical evidence had better be forthcoming. And from Gasser's dedicatory letter to Rheticus it also ought to have been perfectly clear that the weight of this task rested squarely and, for now, exclusively on the shoulders of Rheticus.

In traveling to Italy, Rheticus was looking to acquire not only new mathematical knowledge but also some borrowed prestige. He was claiming—not insincerely—to trade in astrology while aiming to bolster a new cosmology. When he arrived in Milan in the autumn of 1545, he carried with him genitures—horoscopes—that he had collected back in Germany and knew Cardano would be eager to see. Cardano saw the arrival of Rheticus as fortunate, even providential: "Just as I was considering adding some more horoscopes, by chance [*forte fortuna*] Georg Joachim Rheticus, a cultured man and an expert in mathematics, came from Germany to Italy. Immediately, this gentleman . . . offered me some genitures of famous men which he had with him, [including] those of Vesalius, Regiomontanus, Cornelius Agrippa . . . and Osiander."[8]

The name of Osiander on this list of important people, however, would have reminded Rheticus of something uncomfortable: Cardano enjoyed a flourishing relationship with the same Nuremberg publishing elite who, two years earlier, had been responsible for "the stain of chicanery" (Giese's phrase) that marred the first appearance of Copernicus's *Revolutions*. Rheticus was aware that in January of 1545, a few months prior to his arrival in Milan, Cardano's great treatise on algebra (*Ars Magna*) had been published by Petreius. Rheticus also knew that it opened with a glowing dedication to none other than Andreas Osiander himself, whom Cardano addressed as "most learned Andreas," and to whom he offered, as he put it, "the everlasting witness of my love toward you."[9] Cardano was an insider in a tight circle of Nurembergers whose service to Copernicus hung under a black cloud. A similar obscurity darkens Cardano's relationship with Rheticus, even though the appearances seem to display great respect and cordiality. In addition to praising Rheticus as "expert in mathematics," "honorable and meticulous," Cardano called him "exceedingly skilled in the motions of the stars" (*syderalium motum peritissimus*).[10]

The real attitudes of Cardano and Rheticus toward each other, however, remain vexing. One historian points out that in the same 1547 work in which Cardano called Rheticus "exceedingly skilled" as an astronomer, he also recorded a conversation the two men held in Milan on March 21, 1546: Rheticus showed Cardano a certain person's horoscope, and Cardano almost magically, to Rheticus's astonishment, was able to interpret it accurately and in detail. Cardano appeared "equally deft—and Rheticus equally dim"—when Cardano then brilliantly interpreted yet another horoscope he was offered. According to this view, Rheticus is portrayed unflatteringly "as Dr. Watson to Cardano's astrological Mr. Holmes."[11]

A contrary, more positive reading of the two men's relationship downplays Cardano's self-aggrandizement at Rheticus's expense in such isolated scenes. It emphasizes instead Rheticus's persistent interest in Cardano's work over the following decades, as well as the fact that Cardano in his own autobiography listed the name of Rheticus ("Giorgio Porro") among those of his friends.[12] Yet even if the two men never "came to blows,"[13] the legacy of Rheticus's pilgrimage to Italy in 1545–46 was one of disappointment and frustration.

At a personal level, this journey may have shattered Rheticus's cosmopolitan image of himself. Despite his family's Lombardian roots, Rheticus perhaps discovered in Italy how German he was.* At the scientific level as well, Rheticus's experience with Cardano was consistently disillusioning. As late as 1554 Rheticus hoped to glean something from that brilliant mind that would bolster a pro-Copernican mathematics. Here, too, disappointment prevailed: "From Italy," he reported to a friend, "I am receiving neither books nor letters. I really would have loved to see Cardano's completed work, for I was hoping it would be of some use to me in grappling with the science of triangles."[14] The contrast with his earlier experience was undeniable. As Rheticus wrote to another friend: Whereas in Varmia he had met that "eminent man Nicolaus Copernicus," from whom he "absorbed and grasped the exceedingly splendid art of astronomy," in Italy he acquired from scientists of "outstanding reputation . . . very little that was of any use in [his] studies."[15]

If Cardano's work made little contribution to Rheticus's research, Cardano himself was even less useful to the larger cause of Copernicanism. A year after Rheticus left Milan, another tome by Cardano on astrology and astronomy emerged from the printing house of Petreius—*Five Small Books* (*Libelli quinque*)—the fourth of which, whether brazenly or coincidentally, in fact bore the same Latin title as Copernicus's 1543 treatise, *De revolutionibus*.

In an appendix headed "Astronomical Aphorisms," Cardano mentions Copernicus only twice, once disapprovingly, and once with reluctant approbation (Copernicus is "not entirely wrong" in asserting that the moon operates differently from the other planets).[16] Yet neither of these references displayed any engagement with or interest in Copernicus's

---

*Writing to Johannes Crato years later, in June of 1554, Rheticus would refer to Cardano as depriving a German, "our Regiomontanus," of honor that was due him (Karl Heinz Burmeister, *Georg Joachim Rhetikus, 1514–1574: Eine Bio-Bibliographie* [Wiesbaden: Guido Pressler, 1967–68], 3:121). Still later, writing to Petrus Ramus in 1568, Rheticus declared, "I am composing a German astronomy for my Germans" (see the appendix, item 11).

overall teaching, nor did any trace of Copernican influence appear elsewhere in this publication, whose substance remained wholly and fundamentally Ptolemaic.

It is an acknowledged truth that failure, as well as success, can impart valuable lessons. Measured against what Rheticus had likely hoped to accomplish—new mathematical insights for himself, greater recognition for Copernicus, new converts to heliocentrism—his visit to Italy in 1545 and 1546 was an unmitigated failure. The powerful lesson to emerge from that journey, however, was the uniqueness, the singular quality, of Rheticus's relationship with Copernicus.

By the time he prepared to return northward in the autumn of 1546, Rheticus had realized beyond any shadow of doubt that Cardano was no Copernicus. Apart from the fact that both men were renowned as medical practitioners, they offered a stark study in contrasts. Cardano was as flamboyant as Copernicus was reserved, and as self-seeking as Copernicus was generous. Whereas in Copernicus Rheticus had met with patience and fatherly equanimity combined with intellectual brilliance, in Cardano he found only what a later writer would call "spiritually deranged genius," or, in the words of another, a "fractured soul."[17]

Yet the failure of this journey eventually deepened Rheticus's devotion to Copernicus, as well as intensifying his awareness—as Gasser had already so strongly indicated—that he must forge his own implements for cultivating the Copernican project.

CHAPTER 10

SOMETHING ABOUT AN EVIL SPIRIT

On July 23, 1546, a few months before Rheticus left Italy for good, Leipzig's dean of arts, Blasius Thammöller, wrote a polite but firm letter reminding him that he had already been absent for a full year. The letter contained the ominous warning that, unless he returned soon, "those of ill will" might no longer be prevented from disparaging him.[1]

In keeping with his somewhat clumsy way of dealing with institutions, Rheticus hesitated for a time to make any response at all. Then in the autumn, too late for him to comply with the faculty's request for his return, he wrote to Caspar Borner, a senior figure at the university. The letter offered Borner a fulsome account of the valuable work Rheticus had been engaged in. Rheticus went on, however, to complain of his difficult financial circumstances and to ask for an increase in his academic salary.[2]

It was an astonishing and highly impolitic request. Not only would it be turned down by the faculty; it also provoked suspicion concerning his motivation, even his honesty. For in defiance of the university's express instructions, he would persist in his absence for another full two years. From the university's point of view, his behavior appeared inexplicable.

Some stories about what happened to Rheticus between late autumn 1546 and midspring 1547 took on a life of their own, though their common theme is insanity. Luca Gaurico, the pope's astrologer and a rival of Cardano's, whom Rheticus may have met during his time in Italy, offered a concise and, had it been true, unimpeachable excuse for Rheticus's failure to return to Leipzig: After returning home from Italy (Gaurico wrote), Rheticus was overcome by madness and, in April of 1547, died.[3] One other story pointed to a period of temporary madness brought on by scientific frustration.* Yet no solid evidence suggests that astronomy played any direct role in Rheticus's breakdown, even if consciousness of the futility of his Italian pilgrimage contributed in a general way to his condition.

Nevertheless, that Rheticus truly suffered a crisis of the soul is attested in a long letter sent by his trusted friend Caspar Brusch to Joachim Camerarius in late August of 1547. Camerarius, one of the two or three most learned men in all of Protestant Germany, had been instrumental in calling Rheticus to the University of Leipzig in 1542—at some risk, given Rheticus's known record of repeated absenteeism during his employment at Wittenberg. So Camerarius had a personal as well as an institutional stake in having Rheticus return or at least explain himself.

As for Brusch, his friendship with Rheticus stretched back to their years in Wittenberg in the 1530s, when both had belonged to the circle of poets and pub crawlers, a number of them originally from Rhaetia, that included the notorious "shit poet" Lemnius. Later, the two friends had met up on their way to Leipzig, where they arrived and roomed together. An admirer of his friend's scientific achievements, Brusch penned a poem praising Rheticus as a man

---

*Johannes Kepler, in what may have been an act of pure transference, wrote in the dedication of his *New Astronomy* that "Georg Joachim Rheticus (a disciple of Copernicus . . .), when he was brought up short in amazement by the motion of Mars, . . . fled to the oracle of his familiar Genius, . . . whereupon that stern patron . . . caught the importunate inquirer by the hair and . . . threw him down, flattening him on the paved floor, adding the reply: 'This is the motion of Mars.' The story is a bad thing . . . It is nevertheless not unbelievable that Rheticus himself, when his speculations were not succeeding and his spirit was in turmoil, leapt up in fury and pounded his head against the wall" (p. 32).

whose high genius
Has now the stars subdued.[4]

After leaving Leipzig, Brusch had become a schoolmaster back in his home province, in Lindau, and Rheticus turned to him for help during the dark days of late 1546 following his return from Italy.

Having sheltered Rheticus, therefore, the loyal Brusch faced the challenge of offering Camerarius a convincing if belated factual explanation that would cast some positive light on Rheticus's apparently blameworthy behavior. His narrative, brightened by flashes of poetry and touches of drama, offered an extensive firsthand account of a crucial scene in the life of Rheticus.[5] Brusch mentioned Camerarius's request for some report "concerning Joachim's health" and

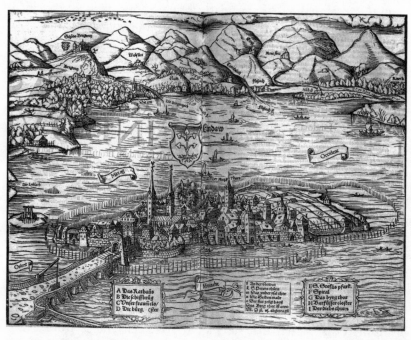

Sebastian Münster's depiction of the island city of Lindau (Cosmography, 1550).

apologized for not having provided it sooner. As he also acknowledged, Rheticus himself was worried that rumors circulating among commercial travelers—"something about an evil spirit"—might damage his reputation.

What Brusch offered in place of such rumors was a slowly unfolding picture of conversion, a watershed that divided Rheticus's life spiritually between a "before" and an "after." Referring to Rheticus's previous combination of irreligiosity and interest in astrology, Brusch wrote that, before this experience, Rheticus had been one

Who, of heaven, pondered naught but signs, while he,
On earth, in Heavenly Father scarce believed.

In Lindau, however, while Rheticus languished in bed for almost five months, Brusch visited him regularly, bringing him devotional writings by reformers such as Luther and Melanchthon, which "he read and reread so diligently that in the end he knew them through and through." At times, "with a full heart and most fervent vows, indeed often in tears, he would call upon the Son of God, awaiting deliverance from him alone." As his spiritual health improved, Rheticus, sharing the lessons he had learned, "vehemently exhorted us to shun the world's and humankind's egregious complacency, whereby we live day by day disbelieving that there are evil spirits or any avenging furies called forth by our wickedness."

In his letter to Camerarius, Brusch emphasized that the change in Rheticus involved a genuine Protestant-style conversion of the heart, not something merely external. Although his Catholic mother and stepfather "pressed [Rheticus] fervently to make a pilgrimage to the renowned shrine of St. Eustatius in Alsace (where the papists deliver many from demon-possession) . . . he would have none of this. It was his conviction instead that he should seek deliverance from Christ alone."

Yet there would appear to be more to the wickedness from which Rheticus sought deliverance than merely a beer-soaked academic lifestyle such as he had earlier pursued in Leipzig and Wittenberg. Surely something more recent, more serious, was afflicting the conscience of Rheticus. In any case he experienced his demoniac attacks as punishment for whatever that wickedness might have been, whether it was associated with Cardano or some other aspect of his Italian sojourn.*

Lest anyone suspect that Brusch was simply making weak excuses for Rheticus's prolonged convalescence, his account is corroborated by an independent source. Brusch had made an excursion in early May of 1547 to visit colleagues in St. Gallen, south of Lake Constance, carrying with him letters of greeting that Rheticus wanted sent on farther south to friends in Zurich. Rheticus himself remained in Lindau for a short while but then in mid-May, just as Brusch was returning home, left for the city of Constance, on the southwestern shore of the lake.[6]

There Rheticus managed to establish close ties with men of highest station, in particular the theologian Ambrosius Blaurer (1492–1564). Blaurer had pioneered the Reformation in Constance and retained close ties with Wittenberg. On December 2, 1547, some months after Brusch's account to Camerarius regarding Rheticus's illness, Blaurer would write a letter to the Zurich reformer Heinrich Bullinger that included a warm recommendation of Rheticus. Like Brusch, Blaurer presented Rheticus as almost but not quite fully recovered ("Satan continues to tempt him," he asserted); nevertheless, he considered the case hopeful, for Rheticus's behavior in Constance had been "above reproach."

---

*Given the strong association in this period of homosexual practice with Italy—for example, the German verb to bugger was florenzen ("to florence," by association with the city of that name)—it is not impossible that Rheticus had undergone some sort of "coming out" experience during his time in Milan. For more background on related issues (such as the common association of sodomy and heresy), see Helmut Puff, *Sodomy in Reformation Germany and Switzerland, 1400–1600* (Chicago: University of Chicago Press, 2003).

As Blaurer attested, Rheticus "is an outstanding mathematician and taught usefully here for more than three months."[7]

That teaching, however, had ended by late summer 1547, and by autumn Rheticus was again on the move. On November 10, from Bregenz, he wrote to Camerarius in Leipzig with yet more foot-dragging on the matter of his long absence from Leipzig. Acknowledging receipt of letters from his mentor and rector, Rheticus professed: "I am ready to lay all other things aside and return to you . . . If you seriously call upon me to return, I shall certainly do so; but if you permit me to spend the winter in Zurich, I shall do that. To you I am bound, and I remain at your command."[8]

Before Camerarius could issue any command to the contrary, Rheticus did indeed make for Zurich—as he apparently intended to do all along. The way was already well paved by Blaurer's commendatory letter to Bullinger. For Rheticus, moving to Zurich meant returning to a place, and some people, he knew and liked. There he hoped (according to Blaurer) "for a time to devote himself to medicine" under the supervision of Conrad Gesner.[9] Gesner, Rheticus's old friend and former fellow pupil in Zurich's Frauenmünster School, had become one of the most famous scholars in northern Europe. He is known today as both the father of zoology and the father of bibliography. Studying with him would offer Rheticus a more comprehensive taste of the sciences, including medicine, than he had so far gained through his studies in mathematics and astronomy. It would also reconnect him with other trusted, upright friends who might smooth his path to physical and spiritual recovery.

Shortly after this move to Zurich, late in 1547, the Leipzig Faculty of Arts sent a sharply worded letter to Rheticus in Bregenz—where they presumed he was still awaiting instructions from the rector—once again urging their truant fellow professor to return. Refusing Rheticus's earlier request for more money, the letter admonished him to be content with his previous salary. Despite its rather cross tone, however, the letter took pains to offer reconciliation—and a resumption of his salary—if only Rheticus would get back to work. "Do come as soon possible," he was urged. "From the very day you set forth to return to us with the aim of re-commencing your work without delay, you shall once more receive your former salary. And if you do this, you will find everyone friendly and affable toward you, and for its part our whole faculty of arts will be highly grateful."[10]

Only a few details survive concerning Rheticus's remaining time in Switzerland from early December 1547 until he finally returned to Leipzig ten months later. He was on the mend, and for this reason his excuses for further delay were sounding more and more labored.

One useful tactic for bolstering his credibility was to present himself as a productive sabbaticant working hard on his research and sharing credit for it with his home institution. In early February 1548, in an encyclopedic folio volume compiled by Gesner, Rheticus published a brief but detailed new proposal for improving the accuracy of an astronomical instrument known as the triquetrum, a large wooden triangular device used for measuring the angular elevations of stars. The dedication of this contribution ("to the teachers and professors of the faculty of arts at the University of Leipzig") suggested how keen Rheticus was at this point to ingratiate himself with his academic colleagues. The professed motivation for the work was Rheticus's perception that "the books of *The Revolutions* of Nicolaus Copernicus have stimulated certain eminent men to observe the motions of the heavenly bodies."[11] In addition to medicine, Rheticus's mind was therefore once again on Copernican astronomy and triangles—a sure sign of returning health.

News of Rheticus's nearly recovered mental and physical state was carried from Zurich back to Constance. Blaurer, in a letter to Bullinger roughly simultaneous with the appearance of Rheticus's new publication, rejoiced at reports "that our Joachim is feeling better. May God utterly destroy the work of the devil!"[12]

INSTRUMENTUM PARALLATICUM SIVE
REGULARUM

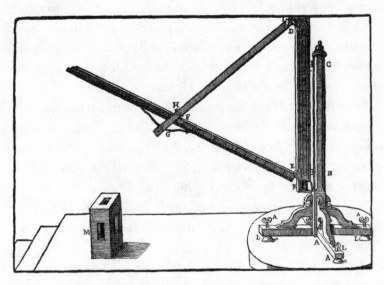

*Tycho Brahe's depiction of Copernicus's own wooden "triquetrum," also known as Ptolemy's rulers (*Astronomiae instauratae mechanica, *1598, 1602).*

On Monday, February 13, Rheticus wrote to the dean and members of the arts faculty in Leipzig concerning his long absence and anticipated return. After the usual pious greetings, he recalled the letter from Thammöller—written a full year and a half earlier—urging him to return to his duties. While acknowledging the "very serious illness" (*gravissumum morbum*) into which he had since fallen, he asserted that the public turmoil (of the Schmalkaldic war),* more than his private ill health, had prevented him from returning from his home region to Leipzig. Nevertheless, he keenly anticipated (he said) the resumption

---

*The Schmalkaldic war, 1546–47, was fought between the Catholic forces of the emperor, Charles V, and the Lutheran princes of the Schmalkaldic league, an alliance that had been formed in 1531 in Schmalkalden.

of his academic duties and hoped, upon his return, that the university would "remember the salary increase" he was promised.*

Rheticus closed the letter by apologizing for yet a further postponement of his return. His doctors were to blame, he wrote, for insisting that he spend some time visiting thermal springs in the Swiss town of Baden, not far from his quarters in Zurich, so that he could "strengthen the nerves of a dislocated foot." He planned to be in Baden at Eastertime and then return to his university.[13]

This was a promise that Rheticus kept, and after making various other brief stops along the way, he arrived back in Leipzig in the late summer of 1548.

---

*Rheticus had apparently misremembered or misconstrued the letter from the dean only two months earlier, in which his request for a pay raise was refused.

CHAPTER 11

LABORS OF HARVEST

Rheticus had been absent from Leipzig for just over three years, slightly longer than he had previously served (October 1542–July 1545). Even though he had been genuinely sick, his Leipzig colleagues may have felt that Rheticus had taken undue advantage of their patience. Even in the faculty's warning letters, however, it was clear that the desire to have Rheticus back at the university was no mere legality. The professors and students of the University of Leipzig sincerely needed him to be there, and they still held him in high academic esteem.

On September 22, 1548, his old friend Paul Eber in Wittenberg sent a letter to a friend in Nuremberg, where Rheticus had stopped on his return journey, expressing pleasure at the reports that indicated Rheticus was returning and in good health. The only punishment Rheticus seems to have faced upon his return to Leipzig was an immediate heavy dose of administration: He found his fellow professors in Leipzig eager to elect him to the position of dean of the Faculty of Arts, and this they did on October 13, 1548.

Rheticus, with recovered health and energy—and having put the maddening Italian journey and its aftermath behind him—now entered upon two of the most intensely productive years of his life. In addition to his administrative duties as dean, he gave lectures in astronomy and mathematics, including his beloved geometry. With a renewed sense of Copernican mission, he wove together these subjects and began to forge what would become modern trigonometry. Rheticus started to realize the inordinate amount of time and extra help that trigonometric tables would demand: seemingly endless calculations to arrive at the correct, precise values for sines, cosines, and the rest that were so necessary for mapping the world, including the heavens. As with any such large, long-term project, this would continuously have to compete with other duties for Rheticus's time.

But it also demanded money. The unrelenting struggle to fund his project would dominate much of the rest of Rheticus's career. His salary as a professor was already handsome by the standards of the day. The University of Leipzig held firm on its decision to give him no raise in pay, although it eventually granted him the nonstipendiary honor of being enrolled as a professor not only in the Faculty of Arts but also in the "higher" Faculty of Theology. This promotion was supported by his old mentor Melanchthon in Wittenberg,[1] as well as by the powerful humanist courtier and nobleman Georg von Komerstadt (or Kummerstad).

Like some academics today, Rheticus was torn between specialized research and work that might appeal to a wider audience. In his case, this tension largely corresponded to the connected but divergent fields of astronomy and astrology. For centuries, the terms *astronomy* and *astrology* were almost interchangeable, but by the mid-sixteenth century, the familiar modern contrast was emerging. As one writer explained in a work published by Petreius in 1542, astrology turned not on the determination but on the interpretation of astronomical data: "The astrologer is concerned with the effects, the astronomer with the causes."[2] The common people's interest—then as now—in those supposed effects offered scholars like Cardano, Gasser, Rheticus, and many others the opportunity to publish material for which there was a considerable market and from which there was some potential financial return.

Astronomy and astrology, while divergent, were nevertheless still so closely connected that it is not quite clear—and may not have been perfectly clear to Rheticus—how far his work on calendars and ephemerides was a logical extension of his astronomy, and how far it was mere popular exploitation driven by a desire to make money.* Even if the profit motive drove such publications, however, Rheticus was adamant that everything was being done for the sake of future work—all of which was essentially an outgrowth of his inherited Copernican legacy.

This conviction is apparent in a letter that Rheticus sent to Copernicus's best friend, Bishop Tiedemann Giese, in Löbau, on October 14, 1549, in which he announced that his calendar and what he called "my Euclid" would be appearing soon. Of the former, he wrote, "Five thousand copies are appearing, all of which I shall send to Bernard Tuyl in Danzig . . . I do hope they sell well, not so that I myself may make a profit, but so that I might fund my projects for the coming year."[3]

In keeping with sixteenth-century practices, these five thousand copies of Rheticus's calendar were shipped as unfolded, unbound sheets to Prussia, where Tuyl would fold them, bind them, and market them according to his best assessment of the preferences of local buyers.[†] Because such works catered to the mass market, they had to be sold very cheaply and were thus also cheaply produced. Precisely because their subject matter was ephemeral, no one cared that they were not made to last. Copies unsold by early in the relevant year were recycled by the printer or bookseller for the value of their paper content.

---

*Ephemerides were almanacs tabulating the positions of heavenly bodies for a particular period. One famous example of this genre was published by Regiomontanus in 1474, a copy of which accompanied Columbus on his fourth voyage to America, where it helped him predict the lunar eclipse of February 29, 1504. See Eli Maor, *Trigonometric Delights* (Princeton: Princeton University Press, 1998), p. 43.

†Five thousand was an impressive number of copies to be sending to only one geographic location. Today, university presses publishing any single scholarly book generally print fewer than a thousand copies for worldwide distribution.

And once that year had expired, copies of calendars and almanacs that had been purchased would play a different role in their owners' out-houses, though not as reading material.

It is no wonder, then, that for many of these cheap, transitory publica-tions, not a single copy survived. Such was the fate of Rheticus's calen-dar for 1550, whose existence is attested to not only in the letter to Giese but also by a court case in which Rheticus became embroiled in October of 1549. Along with popular literature of other ages, sixteenth-century calendars and ephemera achieved part of their appeal through pictures, in this case renderings of astrological signs and their associated animals. These pictures were printed from metal engravings, and Rheticus had engaged a goldsmith named Lorenz Albrecht to make 134 such copper plates. In October, as the work was about to ap-pear, Albrecht demanded payment of six groschen per plate, for a total of fifty florins. It is unknown what the original agreement had been, but Rheticus refused to pay more than twenty-four florins, and Al-brecht took him to court.

This sort of civil proceeding fell within the jurisdiction of the uni-versity. Both Albrecht and Rheticus appeared on October 29, 1549, and it was agreed that an expert third party should assess the value of the work done on the metal plates. Accordingly, on December 18, an independent goldsmith reported to the rector that a fair price for the job would be thirty-four florins. The final meeting of the parties was scheduled for January 9, 1550.

When the time came, though, Rheticus was in his lodgings enter-taining the influential nobleman Christoph von Carlowitz, and he had no intention of neglecting the rich and powerful for the sake of a mere smith. So he sent his famulus to appear in his behalf. Exasperated, the court sent Albrecht home and referred the matter for arbitration to two high-ranking university officials, one of whom was Rheticus's sup-porter and former rector Joachim Camerarius. Albrecht would have none of this and appealed the case to the mayor of Leipzig, who in turn appointed his own arbitration committee. The settlement, handed down on January 22, was the same one recommended earlier by the independent goldsmith: thirty-four florins. Rheticus paid at once, and Albrecht dropped his suit.[4]

The details of this episode offer a glimpse into Rheticus's life between 1548 and 1550. It is clear, for example, that he was financing his own publications. A good deal had to be invested in these entrepreneurial projects well in advance of their realizing any profit. Thirty-four florins was a considerable sum—a quarter of Rheticus's annual salary, and this for one single aspect of one publication. It also seems likely that Rheticus, both in his original agreement with Albrecht and in his response to the suit brought against him, could have paid closer attention to detail and saved himself a lot of trouble. In a phrase Giese had once used to describe Copernicus, Rheticus's behavior appeared to be touched with "apathy and carelessness" when it came to matters nonscientific.

The scene in which Rheticus hosted Carlowitz in his lodgings while neglecting the goldsmith Albrecht is a highly consistent part of this overall picture. Albrecht was a nobody, whereas Carlowitz was a nobleman who personified learning as well as power and wealth. A former student of Erasmus, he was a humanist-diplomat in the service of the duke of Saxony and continued to take an active interest in the University of Leipzig, whose reorganization he had overseen after the Reformation. Rheticus delighted in entertaining such potential supporters of his work, which included his burgeoning project on triangles.

It was Carlowitz to whom Rheticus dedicated his scientific edition of Euclid's *Elements* in November of 1549, a publication mentioned in the letter to Giese the previous month as "my Euclid." The *Elements* was the bible of geometry, and for Rheticus to edit this work—including Camerarius's Latin translation—was to announce in a major fashion his humanist allegiance. The book's dedication itself was long-winded, almost pretentious. Rheticus came dangerously close to saying that geometry is the meaning of life, or in any case the foundation of all human learning. Particularly in the case of astronomy, as he declared, "the observation of the heavens and the explanation of its motions together with the revolutions of the stars, . . . these truly can neither exist nor endure without the science of geometry." He likewise praised geometry as essential to architecture, civil engineering, painting, sculpture, and music, for without it there is "no dimensional, no radial, no harmonic proportion, all of which geometry alone reveals." Finally, geometry is to be praised not only for its usefulness but also—indeed much more—"for its honor, its beauty, and its delight."[5]

Despite all the positive emphasis on the value of geometry, the letter to Carlowitz betrayed a peculiar anxiety. Part of this took the form of the standard humanist "barbarians-at-the-gates" rhetoric. This time it was not the Turks who were threatening civilization and the educational enterprise, but rather those ignorant of geometry and of the benefits it bestows—in particular, those who do not give geometers the gratitude and assistance they deserve.

Rheticus continued that mathematicians, although they create wealth, do not seek it for themselves. Instead, they are like well-bred hounds which, after "strenuously and fearlessly chasing down hares or other beasts," do not devour what they have killed but render them up for the benefit of others.[6] In this way, in a letter intended to attract patronage, Rheticus was able to imply that, although his efforts deserved financial support, he was not seeking money for his own selfish ends. He made the case with some elegance yet also with some of the nervousness one feels when holding out one's hand while wanting not to appear grasping.

A year later, in November of 1550, Rheticus made another indirect plea for patronage, this time in a preface addressed to another noble benefactor of the University of Leipzig, Georg von Komerstadt

(1498–1559). Both Komerstadt and Carlowitz were in the service of Prince-Elector Moritz of Saxony, though the two courtiers were political rivals, and Rheticus apparently decided that the prudent course was to cultivate good relations with both of them. Like the dedication written to Carlowitz, the letter to Komerstadt sought a balance between bold claims about the value of Rheticus's work and a more humble picture of the servant-scientist.

The title of the offered work was *Ephemerides*, whose meaning, along with an unambiguous declaration of the author's principal loyalty, was displayed in the subtitle: *A Setting Forth of the Daily Position of the Stars . . . by Georg Joachim Rheticus according to the theory . . . of his teacher Nicolaus Copernicus of Toruń*. This publication had two prefaces: the letter to Komerstadt and a more general address by the author to the reader. The former opened with a passionate, audacious analogy between Rheticus's scientific struggles and the most famous of all epic wars, in which he likened himself to the Greeks and Trojans "waging war so long over a woman whose beauty was like that of the immortal gods." This image not only exalted the beauty and significance of the object of the struggle—mathematics, and indirectly astronomy—but also drew attention to Rheticus and his need for time and support so that he could grasp heroically the opportunity that stood before him.

Rheticus pressed his analogy toward a Copernican climax. His struggle, he implied, was an extension of his northward pilgrimage to Varmia and of the efforts he had already expended in getting *The Revolutions* published. Mingling humility and ambition, he claimed "no cleverness or wisdom, but only things for which in general nobody envies another—sweat, late nights, trouble, and journeys." Accordingly, he continued, "I searched for someone who could interpret the heavens and instruct me concerning the stars . . . In Prussia I learned and grasped the splendid art of astronomy while staying with that most distinguished man Nicolaus Copernicus. And to work out all these things, to unfold them and to embellish them, is more than the life or the efforts of any one person can accomplish."

Rheticus closed his letter by asking Komerstadt for his fatherly protection. This request makes sense, though, only if Rheticus is seen as defending not just a small book of predictions but a larger, ongoing Copernican project—something indeed opposed by a large array of hostile forces: "There are those who, knowing nothing, will turn away in stupidity; others, through their exceeding knowledge, will rashly carp at some bit they have seized upon; still others will ridicule our entire work on account of its novelty. However, there is nothing one can write by way of defense against all these. Only a bulwark of support will suffice. What will convince them is not disputation but refutation."

In the *Ephemerides'* second preface, "The Author to the Reader," Rheticus likewise boldly showed his Copernican colors. Praising Copernicus as the one "whose hand as it were advanced the machinery of this world," Rheticus warmly recollected: "Driven by youthful curiosity . . . I longed to enter as it were into the inner sanctum of the stars. Consequently, in the course of this research I sometimes became downright quarrelsome with the best and greatest of men, Copernicus. But still he would take delight in the honest desire of my mind, and with a gentle hand he continued to discipline and encourage me."[7]

This vignette of the overeager, even petulant youth being patiently nurtured to adulthood by the wise, fatherly Copernicus is one of the most personally vivid scenes that survive from the life of Rheticus. By sharing it with his reader, Rheticus pointed to the present as well as the past, to his ongoing, undiminished filial loyalty to his teacher—loyalty that was scientific as well as personal. As he indicated in various ways, his *Ephemerides* adhered closely to the theories of his mentor: In the whole work, he declared, "I have not wanted to backslide from Copernican teaching, not even by a finger's width."

* * *

CANON
D OCTRINAE
TRIANGVLORVM.

NVNC PRIMVM A GEOR,
GIO IOACHIMO RHETICO, IN LVCEM
EDITVS, CVM PRIVILEGIO IMPERIALI,
Ne quis hæcíntra decennium, quacunǫ forma
ac compofitione, edere, neue fibi uendicare
aut operibus fuis inferere aufit.

LIPSIAE
EX OFFICINA VVOLPHGAN
GI GVNTERI.

ANNO
M. D. LI.

Title page of the Canon of the Science of Triangles *(Leipzig, 1551),*
featuring Rheticus's trademark obelisk.

The greatest, most scientifically significant work that Rheticus produced during this remarkable period was a small pamphlet titled *Canon of the Science of Triangles*, which appeared early in 1551.* It contained

---

*The term *canon* derives from the Greek name for measuring rod and came to mean "law or code." In music, a canon (or monochord) was originally a Pythagorean device—a simple stringed instrument—for measuring and demonstrating musical consonances. Church officials (like Copernicus) were sometimes called canons because they lived according to a certain code. In mathematics, a canon is a code or a set of instructions for figuring things out, for making calculations.

the first six-function trigonometric tables ever published, followed by a dialogue extolling the usefulness of trigonometry—though the term *trigonometry* had not yet been coined. Such tables permitted astronomers and geographers to make quick calculations deriving distances from measurements of angles. Astronomical theories themselves could not be tested without such increasingly accurate mathematical tools— which permitted the "very precise calculation" that Gemma Frisius had referred to in the 1540s upon receiving the first Copernican report.

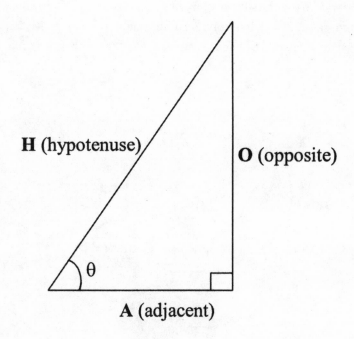

For any right-angle triangle with a given angle θ, six ratios ("functions") can be specified: sine = O/H; cosine = A/H; secant = H/A; tangent = O/A; cosecant = H/O; cotangent = A/O.

The relevance of triangles to astronomy had been illustrated simply in an earlier work by Rheticus, titled *Astronomical Tables*. Triangles can be created by a vertical shadow-stick, or gnomon, by the line of the shadow along a horizontal plate, and by the diagonal line (the hypotenuse) cast by the angle of the sun's rays. In the illustration, the noonday sun casts different shadows at the winter solstice (marked by the symbol for Capricorn), at the equinoxes (Aries and Libra), and at the summer solstice (Cancer). Using trigonometry, one can use the horizontal scale to calculate one's latitude. Accurate calculations for time and latitude can in turn be used as a baseline for other astronomical observations such as those of lunar and planetary position. And the taller the gnomon is, the more accurate one's measurements can be. This partly explains Rheticus's enthusiasm for obelisks, which ap-

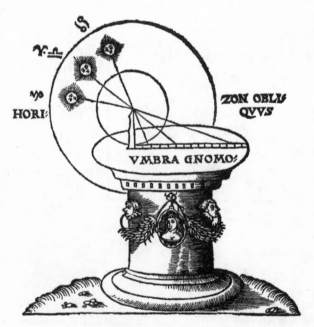

Rheticus's gnomon from Tabulae astronomicae *(Wittenberg, ca. 1542).*

peared on the title pages of most of his works from this period, including the *Canon*.

Finally, appended to the tables of the *Canon* was a dialogue between two characters, Philomathes ("Lover of mathematics, or of learning," Rheticus's own alter ego) and Hospes (a visitor), the first page of which made Rheticus's ongoing Copernican allegiance unambiguous.

> HOSPES: But this Rheticus—What sort of man is he? His name I have heard before, and now I see it written on the front of this little book.
>
> PHILOMATHES: He indeed is the one who is now delivering to us this fruit from the most delightful gardens of Copernicus. For after his recent return from Italy he resolved to impart freely to students of mathematics everything he learned from that excellent old man, as well as everything he has acquired by means of his own effort, perseverance, and devotion.

His loyalty to Copernicus, his contacts with friends and notable people across Germany and beyond, his growing program for honing geometry's usefulness to astronomy—everything that was most important to Rheticus seemed in full flower in late 1550 and early 1551.

Within months of all such signs of success, however, Rheticus would face scandal, exile, and the threatened destruction of his career. In the spring of 1551 a merchant named Hans Meusel (or Meusigen), whose seventeen- or eighteen-year-old son of the same name was a student attending the university, addressed himself to the Leipzig city council, complaining that he had "been inflicted a sudden, outrageous, and unchristian incident shamefully perpetrated by an otherwise reputable member of the university, Joachim Rheticus by name." Meusel declared that "by means of fraud and cunning" Rheticus had not only robbed him and his son of their honor and reputation, but also "put his own life in peril, for which he must indeed surrender his own neck." In particular, Meusel charged, "[Rheticus] with great cunning did lure my son, a minor child, to himself; did . . . ply him with strong drink, until he was inebriated; and finally did with violence overcome him and practice upon him the shameful and cruel vice of sodomy."[8]

This charge spelled the end of Rheticus's career in Leipzig. He could not wait for an opportunity to challenge the details of the accusation. More than his career—his very life, as Meusel correctly stated— was at risk. In April of 1551, just before the end of the winter semester, Rheticus fled the university and the city with sudden urgency.*

---

*Jakob Kroeger, a student in Leipzig, is reported to have written in the margin of a book he was reading that Rheticus had to depart on account of "sodomitical and Italian misdeeds" (*sodomitica et Italica peccata*); cited in Hermann Kesten, *Copernicus and His World* (London: Secker & Warburg, 1946), p. 305.

CHAPTER 12

THINGS LEFT BEHIND

In the sixteenth century universities had legal jurisdictions parallel to those of the cities or towns in which they were located. Some even operated their own student prisons. This division of authority virtually guaranteed frequent "town vs. gown" tensions that were legal as well as cultural. And it meant that either the city or the university could at various times claim—or not claim—jurisdiction for political or other reasons.

When Hans Meusel charged Rheticus with sodomizing his son, he did so before the civic authorities—most likely because he doubted the impartiality of an academic court in a case in which the accused was himself an academic and the plaintiff a townsman. The misdeed was expressly alleged to have been committed, however, on university property and by one member of the university against another. It is not too surprising that on May 5, 1551, the Leipzig city council transferred the case to the university with the request that it undertake criminal proceedings against Rheticus, to be accompanied by a confiscation of his personal property.[1]

For its part, the university did little to dispel Meusel's misgivings about its impartiality, although the rector, Heinrich Cordes, did act immediately to put the case before the academic court, which consisted of himself as president along with four other professors. One of these was Joachim Camerarius, who had hired Rheticus in the first place.*

On May 6 the plaintiff Hans Meusel appeared before the court to demand that Rheticus undergo a full criminal trial. Under Article 116 of the 1532 criminal code of Emperor Charles V, Rheticus thus potentially faced capital charges: "Any person who commits uncleanness with a beast, or man with man, woman with woman, these have forfeited their life, and such a one shall, according to the common usage, be delivered from life to death by fire."[2]

A hearing was postponed to a future date because of the absence of a high-ranking imperial counsel named Ludwig Fachs. A further incentive for this postponement was the rumor that Rheticus himself would soon return, for someone reported that he had indicated such an intention in one of his last lectures. He did not return, and once Fachs arrived back in Leipzig on May 21, the court was obliged to act, though it did so slowly. When, on May 24, Meusel appeared before the rector, he was told that the members of the academic court were too busy with other matters to meet at once. Meusel appeared again four days later, however, and Cordes agreed that every effort should be made to locate Rheticus in Prague, where after fleeing Leipzig he had gone to study medicine at Charles University, and that he should be formally presented with the charges.

At some point during this process Meusel realized that the University of Leipzig was expecting him to bear all the expenses that arose from the case. On May 30 the plaintiff appeared yet again, this time to petition forcefully for exemption from the cost of tracking down the accused and delivering the charges, since he himself had no assurance that Rheticus was even in Prague. The court refused to budge on this

---

*No record survives of Camerarius's personal thoughts concerning the shocking charges being pressed against his former protégé, but years later, in the 1550s and 1560s, Rheticus continued corresponding on familiar terms with both Camerarius and his son.

issue, and on June 2 the university senate approved the decision to deliver charges to Rheticus—at the expense of the plaintiff.

The following day, Rector Cordes sent two sealed letters to Prague, one addressed personally to Rheticus, and the other to the rector of Charles University. The letter to Rheticus did not mince words. After customary greetings, it informed him that Meusel had made "a most serious accusation and criminal complaint against you, namely that . . . in your rented lodgings in the . . . faculty of arts, you did disgracefully abuse his son Johannes in an obscene manner and perform homosexual acts [*masculam . . . venerem*] upon him."[3]

However, once the blunt language was uttered—language that accurately echoed Meusel's original complaint—Cordes softened his tone. Given the university's jurisdiction over Rheticus and the seriousness of the allegations, he wrote, they could not deny Meusel's request that charges be laid. Cordes then ordered Rheticus to appear before the court in the council chamber of the Collegium Paulinum in Leipzig within ninety days to hear the charges and defend himself against them—*or* to send "a suitable, adequately informed representative."[4]

The letter to the rector of Charles University was more formal in tone and much less precise in content. After numerous polite words of greeting, it stated that "the provident Mr. Meusel" had petitioned the university to bring charges against one of its members, "the venerable man Master Georg Joachim Rheticus de Porris," and that "we may not deny his request." No details of the charges were mentioned in the letter, which requested only that the summons be properly delivered and that its receipt by Rheticus be confirmed.

The rector of Charles University received this letter on June 12; on June 13 he wrote back stating that the letter containing the charges had already been delivered to Rheticus. However, the rector noted, when he summoned Rheticus and questioned him concerning this matter, he was told that before settling on a course of action Rheticus wished to confer with his friends, because (according to him) Leipzig had given him only a rather short period of time to respond. Rheticus was not about to share the particulars of his case with anyone in Prague. As far as Charles University was concerned, he maintained a demeanor poised between cautious and casual.

Back in Leipzig, once the original ninety-day deadline for response had expired, formal proceedings began against Rheticus. Rheticus himself did not appear, but he had sent a representative. This outraged Meusel's lawyer, Johannes Rappelt, who protested that Rheticus was in contempt, for in a criminal case the accused must appear in person. Rheticus's representative (whom the records do not name) responded simply by laying before the court his letter of authorization and re-asserting its validity. What was to have been a sodomy trial thus quickly devolved into a procedural wrangle.

The rector had blundered badly in informing Rheticus that he could send an attorney to answer the accusations in his behalf. Yet to admit this mistake not only would be embarrassing but could also render the university liable for the costs it had already extracted from Meusel, and other damages as well. Moreover, a new set of charges would have to be sent to Prague, and the whole weary process would start anew.

The court instead attempted the dubious maneuver of questioning the validity of the credentials presented by Rheticus's attorney. Some-one remarked that these were undated, and that their wax seal looked as if it might have been transferred from another letter and improperly ap-plied to this document—as if a lawyer would resort to forgery in order to play the role of defense counsel in a buggery case.

For his part, an increasingly frustrated Rappelt did not want to incur further delay or further expense for his client, nor did he wish to risk being penalized himself for aborting the trial. Accordingly, he played along with the objections to the attorney's credentials, even while per-sisting in his claim that such deputation was in any case inadmissible in a criminal proceeding. His compliance permitted the court to avoid conceding its mistake and instead to adopt a rather lenient position: It would proceed with the trial, but Rheticus's deadline to appear would be extended by one month.

On September 11, 1551, Cordes wrote a new letter to Rheticus in Prague, admitting no fault of his own, indicating the inadequacy of the designated representative, ominously implying contempt on Rheticus's part, but offering him the postponement of his trial date to Monday, October 12, at which time he must appear in person.[5]

If, however, up to this point Rheticus had remained technically innocent of contempt, he now stepped boldly across the line. His attorney returned to Prague bearing the letter from Rector Cordes along with news of the muddled proceedings. Rather than responding to the substance of the rector's missive, Rheticus simply sent the letter back to Leipzig unopened, accompanied by a note to the effect that he had already responded to his summons and therefore did not believe that this document was intended for him. He observed huffily, the letter was not addressed to the "Leipzig Professor of Mathematics."

The University of Leipzig had no choice but to follow through on the process it had set down. A routine summons-to-appear was read on October 18 and again on November 1 and 22, yet Rheticus did not appear. A collective amnesia then seemed to descend upon the court, only to be punctured by the ever-persistent Hans Meusel Sr.

On March 29, 1552, Meusel petitioned the court to act on its commitment—postponed from October 12—to proceed against Rheticus. By now the members of the court had only the foggiest recollection of the case, and the rector ordered his assistant to review court files in hopes of discovering that the case had indeed already been wrapped up. It had not. The court therefore reconvened on April 11 to render its judgment, which had become a foregone conclusion. Rheticus had been in de facto exile for twelve months, and this made it official. He was banished from the University of Leipzig for 101 years.

The extraordinary thing about his sentence was this: "Upon very weighty considerations—for reasons of both public respectability and

[Rheticus's] exceptional talents—the rector agreed there were grounds for not publishing his sentence in the usual manner. Instead, by public authority a written sentence would be sent to Rheticus from the university court, delivered by Andreas Siber, citizen of Leipzig."[6]

This enormous concession effectively placed Rheticus in a kind of identity protection program—except that it was one in which he could retain his name and identity. He was under permanent banishment from Leipzig, yet the thick curtain of silence that the academic court let fall upon his alleged homosexual crime—independent of lingering questions about whether justice was properly served—would permit Rheticus to pursue his subsequent life and career unmolested and free from further prosecution.

If the terms of Rheticus's banishment from Leipzig were significant for his future life, the records pertaining to what he left behind are intriguing for what they reveal about his activities and way of life until that departure.

Rheticus's legal problems were not limited to the criminal charges for which he was banished. When he departed Leipzig in April of 1551, he left behind a series of creditors, who undertook civil proceedings in hopes of recovering at least a portion of their bad loans. For the sake of these creditors, three different inventories were made of Rheticus's possessions: (1) of the effects from his university lodgings (May 20, 1551); (2) of the contents of his workshop-storeroom (December 1, 1551); and (3) of his remaining books and household articles (February 1 and 2, 1552). One tally of Rheticus's total indebtedness upon his departure from Leipzig puts it at 269 florins and two groschen,[7] or almost twice what had been his annual academic salary.

Only the second inventory is known to have survived, and it offers a freeze-frame glimpse of the life of Rheticus at the beginning of 1551. (For the full inventory, see the appendix, item 6.) The furnishings of Rheticus's workshop-storeroom, which the inventory identifies as a vault, were extremely basic, including a table, a bench, a ladder, and some shelves. Other contents included one piece of astronomical equipment (an armillary sphere) and great quantities of blank paper, along with maps and commercial supplies of printed books.[8]

The enormous amount of paper mentioned in the inventory augments the picture of Rheticus as a literary and publishing entrepreneur. In the sixteenth century paper made up a larger proportion of the cost of book publishing than it does today, and an author with a huge supply of it had more leverage in negotiations with a publisher than an author with none. Leipzig was a publishing center and important venue for trade fairs, and paper was in high demand.

In the sixteenth century quantities of paper were measured in various units. One bale equaled ten reams, which equaled two hundred "books," or five thousand sheets. Different publications required different sizes of paper. The main determinant of the size of each printed book—apart from the sheer length of the text to be set—was the number of times each sheet was folded. Folded once, the sheets would become part of a large-format folio book, and each sheet would form two leaves or four pages. Folded twice, the sheet would become four leaves or eight pages in an ordinary-size (quarto) book. Folded three times, the sheet would form eight leaves, sixteen pages, in a small-format octavo book.*

When the quantities were added up, Rheticus's vault contained 76,725 sheets of paper, enough for over three thousand copies of a quarto book of two hundred pages. Or, for the more commercial sort of almanac such as the large one mentioned in the inventory, which comprised only twenty-four pages (twelve leaves, or three sheets folded in quarto), there was enough paper for more than twenty-five thousand copies.

---

*This process is most easily understood by someone willing to take one sheet of A4 or 8 1/2 × 11 inch standard paper and fold it in half three successive times.

By far the most fascinating contents of Rheticus's vault were the materials there that had already been published. The maps suggest that Rheticus was keeping up his interest in mapmaking earlier shared with Copernicus and Heinrich Zell in Varmia—the interest formally discussed in his *Chorography*.*

The books listed in the inventory also represent vivid connections with other dimensions of Rheticus's career. The one titled *Astronomical Tables* may have been a new edition of a work of the same name by Rheticus that had already appeared three times in Wittenberg in the early to mid-1540s. Because it was a small reference work for use by students, it is likely that the *Tables* would have been periodically revised and reprinted.

The *Ephemerides* mentioned in the inventory was a copy of the same work Rheticus had recently dedicated to Komerstadt, the one in which he had called Copernicus "the best and greatest of men." And the *Canon* listed there was Rheticus's groundbreaking work on triangles (1551)—the first tables of all six trigonometric functions ever published—to which was appended the dialogue announcing that Rheticus was delivering "fruit from the most delightful gardens of Copernicus." That the *Canon of the Science of Triangles* appeared in the vault in huge numbers—1,450 copies—and that it had been published by Wolfgang Günter in Leipzig only weeks before its author fled the city, suggests that Rheticus's vault might have contained virtually the entire print run of this vital work.

The last item mentioned in the inventory was a short book on how one should prepare to face death. Among the books listed, it is the only one not written by Rheticus himself. Its title, *On Death*, together with the implied number of pages (nine and a quarter sheets, hence an

---

*Zell had stayed on in Prussia and filled a dual role as lecturer at the new University of Königsberg and as librarian to Duke Albrecht. In 1550 he published two versions of a map of Germany, one in Latin and one in German, named in the inventory separately as Germania and Deutschland.

octavo book of 74 leaves or 148 pages), together with its time frame (all of the other items were recently published), makes it certain that this was a printed sermon by the southwest German reformer and spiritual leader Johannes Brenz.[9] Its full title was *How One Should Prepare Oneself in a Christian Manner to Face Death.*[10]

The inventory leads directly to a recognition of just how complex and conflicted the life and character of Rheticus were. As a whole, it offers a picture of a man devoted to a program of learned—though also to a degree popular—dissemination of scientific knowledge. And yet the very circumstances that made the inventory necessary are a reminder that the tapestry of Rheticus's life was interwoven with a pattern of financial exigency occasioned both by the scope of his ambitions and, on the negative side of the moral ledger, by the excesses of his personal habits.

Beyond the specific charge that he criminally plied with drink and sexually abused one of his students on university property, there are other indications. On the list of Rheticus's creditors in May 1551 stood the name of one Georg Helfereich, whom Rheticus owed ten talers (or about eleven and a half florins) "for wine."[11] This amount corresponded to a twelfth of Rheticus's annual salary at the University of Leipzig.

It is impossible to say how far Rheticus's moral challenges were related to drink, pederastic sex, money, or something else. What is not in doubt is that he and a number of men close to him saw his struggle as profoundly spiritual. That was the explicit interpretation put forward by his friend Caspar Brusch scarcely three years before Rheticus's exile from Leipzig.

In Lindau, according to Brusch's letter to Camerarius in late August 1547, during a period of intense spiritual struggle Rheticus had diligently studied "the devotional writings of Luther, Melanchthon, and Cruciger"; then, "on particular occasions, with a full heart and most fervent vows, indeed often in tears, he would call upon the Son of God, awaiting deliverance from him alone." Moreover, he frequently "exhorted us to shun the world's and humankind's egregious complacency, whereby we live day by day disbelieving that there are evil spirits . . . called forth by our wickedness—lacking all fear of God, and not hesitating to do just whatever we please." In short, Rheticus's "conviction" was that "he should seek deliverance from Christ alone, the Son of God, . . . who alone was born to destroy the works of the devil." (For the full text of Brusch's letter, see the appendix, item 5.)

There is no reason to suppose that Brusch had unduly embellished this scene. Although Rheticus's piety did not appear consistently at every stage of his life, it made definite appearances, as Brusch's letter attests; it would do so again in Kraków, years later, in correspondence with his most godly Wittenberg friend Paul Eber. It is the pattern of fervent but arrested piety that offers a third dimension to the shadowy portrait of Rheticus cast by the inventory of his vault. He was driven to publish and to produce within the realm of science. He was driven from Leipzig by his own real or alleged moral failings. And he was driven to seek deliverance in the language and the experience of Christian reconciliation with God.

With this third dimension of Rheticus's portrait, the presence of Brenz's *On Death* among the items in the vault is strikingly consistent. As indicated in Brusch's letter, part of Rheticus's exercise of piety involved not only immersing himself in the leading Protestant devotional writings but also exhorting others to attend to the truths conveyed by those writings. Brenz's *On Death* harmonized with this pattern and used exactly the kind of spiritual, even apocalyptic, vocabulary employed in Brusch's account of Rheticus's conversion.

In his printed sermon *On Death*, Brenz summarized the great divide separating those whose hope is in Christ and those who have turned away.

> Why should a man be dismayed in the face of death when, for him, there shall be no more death, but rather death be transformed, through

Christ, into a rest and a sweet sleep, yea, into a way, a road, and a door of redemption leading to life immortal? . . . Therefore, let those fear death who are not buried with Christ by faith and baptism, who have not laid their sins upon him, who may not cry to the Lord as to a father, who have not received the Holy Ghost, who take no pleasure in righteousness, who have spurned or despised the Gospel, who after this life are fit for nothing but the pains of everlasting hell.[12]

In most literature of this kind, two thematic words sum up the response of the "observer" to the apocalyptic or mortal scene being presented, depending on where he or she stands relative to the great gulf that Brenz depicts. Not surprisingly, these words are *Schrecken* (horror) and *Trost* (comfort). And of course such scenes are intended not simply to describe different individuals' fates but to influence them— to exhort readers not to neglect fundamental issues of life and destiny. They are aimed at urging people—in the words of Rheticus as quoted by Brusch—to "shun . . . complacency."

For Rheticus, the exercise of piety meant delivering this message to others by means of books like Brenz's—but also, in highly compact form, by means of his own publications. And so for readers of the "large" *Almanac* listed in the inventory, Rheticus placed on its title page a stunning précis of Brenz's apocalyptic sermon. The woodcut is of Death, his chariot, and his scythe mowing down all who stand before it; the poem, with its implicit appeal to readers to shun this fate while there is still time, refers directly to the picture.

> *Recoil in horror at this scene—*
> *Yet only if your life has been*
> *Devoted but to vanities,*
> *Heedless of God's own verities:*
> *His Word, His holy will revealed,*
> *His Christ, who our redemption sealed,*
> *Who saves from sin and death and devil,*

Delivers us from every evil;
Our everlasting comfort he—
Praised be his name eternally.

While there may be no way of fully harmonizing Rheticus's disparate concerns and traits of character, it would seem that Rheticus the fervent if conflicted preacher must be kept in dynamic suspension with Rheticus the adventurer, Rheticus the Copernican mathematician, and, starting in 1551, Rheticus the exile. His life and possessions in Leipzig were things he had now permanently left behind.

Prognosticon oder Prac
tica Deutsch / auff das jar Christi
M. D. L I. gestellet / durch Georgium Joachimum Rheticum / Mathematicum zu Leipzig.

Mit befreiung Key. Maiest. vnd Königlicher Wird zu Polen / nicht nach zu drucken.

Dis bild erschrecklich denen ist	Vnd I E S V Christ den ewigen hort,
Die ihren vleis zu aller frist,	Der vns vom Teuffel, Sünd vnd Todt
Legen auff dis lebens pracht	Erlöset vnd hilfft aus aller not,
Vnd geben weiter gar nicht acht,	Vnd tröstet vns inn ewigkeit
Auff Gott des Herren will vñ wort	Dem sey lob, ehr, zu aller zeit.

Title page of Rheticus's Prognostication *for 1551, with ominous graphic and poem.*

CHAPTER 13

MEDICINE IS NOT LIKE GEOMETRY

Exiled from Leipzig—de facto in 1551, and then officially a year later—Rheticus had left behind a good academic salary along with all prospects of patronage and of income from sales of paper or books. Like anyone declaring bankruptcy, he was not only relieved of his debts but at the same time shorn of his credit. He somehow had to begin anew.

What Rheticus did between the time he left Leipzig in April of 1551 and May of 1554, when he sent a friend a letter from Kraków, is somewhat unclear. What is clear is that he spent much of this period acquiring the necessary qualifications to become a doctor.

There was logic in this; his own father had been a doctor, as had Copernicus. So was Rheticus's mentor Achilles Gasser, and likewise his school friend Conrad Gesner, with whom he had studied medicine for a short while in Zurich not many years earlier. If a man wanted a potentially lucrative career that would build on his existing astronomical and astrological expertise yet not tie him to any particular location or institution, medicine seemed to be just the thing.

Arriving in Prague in the spring of 1551, Rheticus had devoted himself to medical training, though no record remains of whom he studied with or what qualification he attained. Despite this definite pursuit of medicine, it is evident that Rheticus never quite gave up thinking of himself principally as a mathematician. On the contrary, the picture that emerges is one of a man busily acquiring and practicing medical skills but always with hopes—realistic or not—of creating greater opportunities for himself to study stars and triangles.

During late 1553 or early 1554, Rheticus was offered a new professorship of mathematics at the renowned University of Vienna, where apparently no one was aware of his banishment from Leipzig or the reasons behind it. That university's records for 1554 list Rheticus as the holder of the third chair of mathematics in the faculty of arts.[1] Whether Rheticus was simply not interested in ranking third among three professors, or whether, as is more likely, he was weary of academic institutions and their inadequate salaries, he decided to turn down the job.

Paul Fabricius, the man who held the second professorship of mathematics at Vienna, wrote Rheticus a long poem in 1554 lamenting that decision. One of the main themes of Fabricius's verses was the tension between the monetary riches offered by medicine and the rewards of a higher kind promised by mathematics. Like the true believer he was, Fabricius extolled mathematics over medicine for its intrinsic worth and beauty—and warned that Rheticus might overburden himself if he tried to serve both arts.

And say you, mathematics knows no riches?
This I concede (not wanting to speak false),
Yet urge you still to ponder this my judgment:
Both arts indeed have value, deserve praise.
One art gives riches; the other art gives none.
But of the arts are riches the true measure?

Both arts are lovely; both are virtuous.
'Tis doubtful which deserves the greater praise.
The one is beautiful; the other richer.
I care for one, am careworn by the other.

Rheticus must have read his friend's verses with immense pleasure. That poem—and Rheticus's connection to Vienna generally—would remain part and parcel of his own and others' conception of him as a mathematician above all. In gratitude, Rheticus responded by sending Fabricius a manuscript excerpt from his still-in-progress "Science of Triangles," calling "this meager little book . . . my greatest treasure . . . All of my other books are nothing but tributaries to this . . . [and] I hope you will keep it for yourself alone, as long as I shall live."[2]

Rheticus's Viennese connections—not only with individuals such as Fabricius but also with the Hapsburg court itself—remained significant as sources of prestige and patronage even beyond his lifetime. Yet Rheticus was by nature so optimistic that he failed to appreciate the full urgency of the requirement, as Fabricius wrote, that he "somewhat abstain" from medicine in order to pursue the "sacred sweetness" of mathematics.[3]

The other place Rheticus visited for a time in the winter of 1553–54 was Vratislavia in Silesia.* There he continued pursuing his study of medicine, assisted by his own former student Johannes Crato (1519–85), at whose M.A. graduation in Wittenberg in April 1542 Rheticus had presided as dean of arts. Crato had recently returned to Vratislavia, his birthplace, where he served as town doctor. His meteoric medical career would later include service as personal physician to Emperors Ferdinand I, Maximilian II, and Rudolf II. As for Rheticus, he soon moved on from Vratislavia eastward to Kraków, which would be his residence for almost the next two decades.

The earliest extant letters written by Rheticus from Kraków, then the royal capital of Poland, were to Crato in Vratislavia. It is obvious from the first of these, dated May 21, 1554, that Rheticus, as Fabricius had warned, was being pulled in two opposing directions. While telling

---

*Today, Wrocław in Poland.

Crato how eager he was to see his new and forthcoming medical publications, Rheticus wrote that he himself was "totally occupied composing the commentary for [his] 'Science of Triangles'": "I am giving this all the time I can possibly divert from medical practice. This work, eagerly awaited by many learned people, is my delight; and perhaps it will occupy my lifetime, indeed my every moment."[4]

Barely a month later, Rheticus wrote once again to Crato. In this letter, although professing to be busy with his "Science of Triangles," he was sounding impatient, as well as lonely and defensive. Lamenting that he had "no one with whom to share [his] interests, neither doctor, nor chemist, nor mathematician," Rheticus went on to grumble concerning "a mighty authority," as he sarcastically called him—one Jan Benedikt, a personal physician to the king making a huge salary of two thousand florins per annum. Despite Rheticus's recurrent financial worries, he informed Crato that he did not intend to curry favor with such mercenary practitioners. Referring to himself in the third person, he declared: "Rheticus pays no honor to ignoramuses . . . Let me stay poor rather than sell myself to asses."[5] His isolation as a newcomer in Kraków, added to his relative poverty, was fueling his resentment of those more successful than he was.

A further thirty days went by, and on July 20, 1554, Rheticus wrote to Crato once more. The letter is a rare snapshot of the range of Rheticus's interests, emotions, attitudes, and activities as he tried, at the age of forty, to settle in to life in Kraków and establish his program of work.

> Mathematical studies here lie in such ruins that, by Hercules, I doubt that anyone can restore them, not even the public professors. At least I have seen nothing of them. They sell their horoscopes for a florin each. I saw one of these drawn up for a prominent person, someone from the royal court, in which there was nothing but annual and monthly revolutions, no directions at all. And they are no better when it comes to geometry.

I did persuade a certain Nolde that he should learn Euclid, which he did in his mediocre way. But now that he might make himself useful, he is avoiding me, so I have decided I no longer want his help. I did not spend all that time and money studying mathematics just so I could badger people to take instruction from me.

I have erected a fifty-foot obelisk in a perfectly level field that the marvelous Mr. Johannes Boner has made available to me for this purpose. By this means, God willing, I shall describe anew the whole sphere of the fixed stars. I could complete this labor faster and more easily if only I had fellow workers. Still, eventually, with God's help, I shall finish my project.[6]

Rheticus fervently longed to achieve scientific progress, and he strongly resented the obstacles that stood in his way. In this cosmopolitan city populated not only by Poles but also by Germans, Jews, and some Italians, there were opportunities for learned men to make a good living. In the sixteenth century mathematicians frequently doubled as astrologers, and Rheticus was no exception. In Kraków in 1554, however, it was conspicuously weak mathematicians who seemed to be profiting from writing horoscopes. Rheticus, to his chagrin, was not. The irony was that such income, even though distasteful, was desperately needed if he hoped to achieve real progress in his mathematical research.

Rheticus also needed trained help, but in 1554 his efforts to train an apprentice produced only disappointment—a single inept student, the lackluster Nolde, whom history knows only for his neglect of the chance Rheticus gave him of helping to extend the frontiers of mathematical knowledge.

Some shafts of light were beginning to appear, however, if not without shadows. Rheticus's hopes soared with the erection of an obelisk—an object that epitomized, among other things, the importance of winning patronage and acquiring the right scientific equipment. These things have always gone together, for the latter, generally being quite expensive, requires the former. The letter reveals, then, that by the summer of 1554 Rheticus had already acquired some patronage, even if he would always need more help, particularly in the form of fellow workers. On balance, good things were starting to happen: the patron, the obelisk, and a growing sense of divine favor, twice explicitly invoked ("God willing," "with God's help").

All of this positive news seemed to resonate with Rheticus's professed reason for pursuing his role as Copernican apostle in Kraków. As he would affirm in his 1557 letter to King Ferdinand, he had devoted himself to astronomy "at the behest of my master Copernicus, whom I revere as not only my teacher but also my father, honoring and ever striving to please him." As Copernicus had pursued his research in Frauenburg, so Rheticus chose Kraków, because it lay on the same meridian. And as he would proudly tell the king: "Here I have erected an obelisk, forty-five Roman feet in height. For in my opinion there is no better astronomical instrument."[7]

For Rheticus, however, feeling optimistic never meant feeling content. In his July 20 letter to Crato he disclosed both old irritations and a new passion. As he had done a month earlier, Rheticus wrote scornfully of the royal physician ("let him do what he likes, I shall never worship that ass"). More seriously, Rheticus uttered a grievance that would be repeated in years to come: a sense of remoteness, in particular the difficulty of obtaining the latest scientific books. So desperate was his need that Rheticus had to ask Crato to send him a copy of his own edition of Euclid.

The new passion, however, was connected with Rheticus's deepening curiosity and experience as a physician. Ironically, this theoretical, scientific interest in medicine added a new threat to prospects for Rheticus's completion of his work on astronomy and trigonometry. It was one thing for the practice of medicine to be taking up so much of his time. If his Copernican work remained his sole scientific passion, he might still hope to complete it. But if medicine began to occupy his mind as well as his time, that was something else altogether.

A glimmer of Rheticus's growing medical curiosity appeared at the end of his July 20 letter to Crato. He begged his friend to send him samples of distillates and oil of vitriol, unavailable in Kraków. He asked him also to share his thoughts on the medical use of oil of antimony. Then, two and a half months later, Rheticus sent Crato another letter thanking him for the oil of vitriol, which he had received. This letter accompanied a favor in return, a small jar of rock salt "from the treasure houses of Poland"[8]—by which Rheticus meant the great salt mines of Wieliczka, near Kraków.

Medicine was occupying his mind more and more, and for Rheticus, medical research was taking a decidedly pharmacological turn.

The most famous figure associated with a new chemical approach to medicine in the sixteenth century was Philippus Aureolus Theophrastus Bombast of Hohenheim, better known as Paracelsus. Paracelsus (1493–1541) was in his own day called the "Luther of medicine" and is today recognized as an early pioneer of the science of toxicology.[9] Partly because he was a (younger) contemporary of Copernicus, his work is sometimes seen as exhibiting parallels with that of the astronomer. These should not be exaggerated. Nevertheless, the work of the two men did raise questions about the adequacy of Aristotelian approaches in physics (Copernicus) and in chemistry and medicine (Paracelsus). Both men looked for harmonies and coherence on a large scale, and both were paid visits by the young Rheticus.

Rheticus himself would claim, in a letter written from Kraków in 1569, that he had met Paracelsus at some point in 1532 and had "spoken with him."[10] The meeting most likely took place somewhere near Lake Constance, arranged by Rheticus's mentor the Lindau doctor Achilles Gasser.[11]

Most of the medical training Rheticus acquired had been of a more traditional, "humoral" sort, drawn from the teachings of Galen. Its pharmacology was based on plants rather than minerals. All of the doctors with whom he enjoyed close relationships, including Gasser, Gesner, and Crato, were non-Paracelsian—indeed anti-Paracelsian—practitioners. As Rheticus's enthusiasm for Paracelsus grew during his Kraków years, he would have to be careful, or at least nonconfrontational, about displaying it in correspondence with his medical friends.

In spite of his former teachers, however, he may have been drawn to Paracelsus because of this reformer's emphasis on correspondences between the individual and the cosmos. Paracelsus's search for deeper causal explanations for anomalies—in this case, of course, not retrograde motions of heavenly bodies but diseases within human bodies—would have attracted Rheticus. This was a genuine link, in principle, between astronomy and medicine. As Paracelsus asserted, "everything which astronomical theory has searched deeply and gravely by aspects, astronomical tables and so forth, . . . this self-same knowledge should be a lesson and teaching to you concerning the bodily firmament."[12]

In the words of one prominent historian of science, "the greatest of scientists have been unifiers."[13] For all his quirkiness and excess of rhetoric, Paracelsus was in spirit another unifier. In following him in the area of medicine, Rheticus was still pursuing his quest to hear the harmony and to see the interconnectedness of all things within the larger universe.

Rheticus's enthusiasm for Paracelsus was displayed in 1558 in an affectionate letter he wrote to his former promoter and rector Joachim Camerarius Sr. The two men had remained in friendly contact despite the elder scholar's full awareness of why Rheticus had been banished from Leipzig. The letter to Camerarius tells of the healing by Paracelsus of someone whom everyone expected to die—and who would have died had he been treated "according to the manner of humoral medicine." As Rheticus recounted, Paracelsus treated the ill man, instead, according to "the true art [of medicine] that God has hidden in nature." He gave his patient wine containing three drops of distillate, and the next day the same man appeared among the guests at Paracelsus's lodgings fully restored, "to the great amazement of all present."[14]

Rheticus thus conveyed not only his ongoing interest in Paracelsus but also his solicitude for a former patron. Camerarius and his wife, as Rheticus had been informed by their mutual friend Crato, were at that time not enjoying very good health. As with anyone who genuinely believes in a particular therapy, Rheticus hinted to his sick friends that they might benefit from trying a similar treatment. Yet even in this letter focusing on medicine, Rheticus could not hide his persistent, even higher regard for another discipline. As he wrote right at the start of the letter, unfortunately "our medicine is not like geometry, which always attains its goal."

Five years later, in February 1563, Rheticus wrote to the son of these elderly friends, Joachim Camerarius Jr., in Nuremberg. In this letter he disparaged Kraków as "the cave of the Cyclops"—Homer's one-eyed monster and the antithesis of all things civilized. Bemoaning the unavailability of books in Kraków, he asked to be sent a list of works written or edited by the senior Camerarius so that he might write away for some of them. But Rheticus also requested that the younger Camerarius help him learn more from Germany about "a new school [of medicine] springing up, whose founder is Theophrastus Paracelsus." As Rheticus continued, "I take great delight in chemistry. Yet I cannot master his whole art of medicine until the appearance of his works on philosophy, astronomy, chemistry, and medicine, upon which the edifice of his art of medicine is founded . . . So please send me by the fastest possible means any publication from this school that might appear."[15]

An ambitious program of mastering the whole art of Paracelsian medicine went well beyond what was necessary for making a handsome living as a medical practitioner. It was clear that Rheticus's interest in this area of research was becoming a full-grown passion. What remained to be seen was whether he could muster the time, energy, and resources to serve two passions, one Paracelsian, the other Copernican.

CHAPTER 14

ANOTHER COPERNICUS

Rheticus's best friend in the 1560s was someone he had actually not seen for well over a decade, the theologian Paul Eber (1511–69). Rheticus and Eber had begun their studies at the same time in Wittenberg and later also received their M.A. degrees together in 1536.

Eber was a man of impressive intelligence, dedication, and saintliness. He came from a poor family who had cut back on their own meager rations of food so that he could attend school in Nuremberg, which in turn groomed him for Wittenberg. Three years older than Rheticus, Eber at the age of twenty-one already had suffering written all over him, almost literally. Eight years earlier he had been thrown from a horse and survived the fall with permanently fragile body and deformed face. Rheticus, never one to overvalue external appearances, perhaps discerned in Eber an outward reflection of his own inner sufferings that dated from late childhood.

The two undergraduates at once struck up a friendship that ended only with Eber's death in 1569—a friendship that flourished despite, or even because of, the two men's profound differences of character. Rheticus recognized in Eber depths of spirituality that he himself never did achieve and possibly never could. The most famous expression of this piety was a hymn, still found in Lutheran hymnals around the world, penned in 1547 while Wittenberg was under military siege and

Eber and two other professors were the lone remaining occupants of the university.

> When in the hour of utmost need
> We know not where to look for aid,
> When days and nights of anxious thought
> Nor help nor comfort yet have brought;
> Then this our comfort is alone,
> That we may meet before Thy throne,
> And cry, O faithful God, to Thee,
> For rescue from our misery.[1]

Such expressions of faith amid misery—like the devotional literature that Rheticus absorbed in Lindau in 1547 and was storing in his vault in Leipzig in 1551—would continue to resonate with him, even if only intermittently. In his letters to Eber in the 1560s he would adopt a noticeably more pious tone than was evident in his other correspondence. Most important, though, Eber remained loyally in contact with him even after the scandal of 1551—at a time when Rheticus was grateful for any faithful friend, especially one of Eber's undoubted character and dignity. Throughout the 1560s Eber carefully filed away the correspondence he received from Rheticus, and these surviving letters offer a sort of scrapbook of Rheticus's thoughts and moods during his first decade in Kraków.

The earliest of these personal letters, dated October 8, 1561,[2] doubled as a letter of reference for the itinerant student named Jodok Rab who delivered it. Rheticus declared him "a decent young fellow" and recommended him to Eber as well as to the mathematician Caspar Peucer, his own former student. The core of the letter focused on Rheticus's and Eber's shared interest in astronomical, astrological, and historical writings. "I know that your meditations on the invisible Heaven do not keep you from considering the visible heavens, and that you are observing the workings of the stars," Rheticus wrote to his theological friend, and encouraged his continued contribution to an ever-expanding work of Protestant historiography known as Carion's *Chronicle*.[3] Do not, he implored his friend, "deprive us, and posterity, of your accomplishments."

These mingled themes of scholarly community and scientific research were complemented by Rheticus's request that Eber send him any newly published works by Ptolemy, Hermes Trismegistus, and Hieronymus Wolf. Such requests for books, like much else in his correspondence, reveal Rheticus's sense of distance from Europe's rich heartland of intellectual activity. As he wrote in the letter, referring to Kraków, "When it comes to mathematics, this place is a wasteland."

The unreliable postal system of the day did nothing to diminish Rheticus's feeling of isolation. Five months later, on March 1, 1562, he wrote again to Eber and admitted he did not know if his previous letter had arrived.[4] To his "fervently cherished friend" Rheticus repeated his interest in the work on Carion's *Chronicle*, but this time revealed plans for a chronology of his own "based on the astronomy of Copernicus, starting with the foundation of the world." Consistent with the dominant pattern of his life after 1539 even this historiographical project would be an outgrowth of Rheticus's dedication to Copernicus.[5] In this work, Rheticus assured his friend, "You will read many things that will amaze you."

Lightening his tone, Rheticus then offered Eber some petulant humor at the expense of Wittenberg publishers and booksellers, whom he accused of disorganization on a cosmic scale—not to mention extreme miserliness in paying freelance editors. Referring to a project he had edited for a publisher named Konrad Rühl,* Rheticus told Eber: "Believe me, I didn't even make enough to pay for the beer I needed to get me through the job. These beasts are so used to getting everything for free."

---

*For Rühl, Rheticus had revised a pair of works, including the *Astronomical Tables*, of which 240 copies were left behind in Rheticus's vault in Leipzig. See the appendix, item 6.

After a renewed appeal to Eber to send him books, Rheticus concluded his letter with a reassertion of what was still his number one priority: "All the money I make here I devote to the production of my canon of the science of triangles. Not a day passes but that I spend a florin on this project—something I have now been doing for six years." One florin represented up to three days' wages for an ordinary university professor. It should have been obvious to Rheticus that his stated plan to "complete the whole thing within the next two years" was unrealistic.

Eight months later Rheticus sent another letter to Eber, one that offered a keyhole glimpse of the theological ferment of the time in Kraków, along with Rheticus's weary response to it.[6] Addressing Eber as "my dearest friend," he lamented, "[Although] medicine readily gives me enough money and prestige . . . I desire nothing so much as one single friend whom I could occasionally talk to." He told Eber he looked forward to seeing his small treatise aimed at settling disputes concerning the sacrament of Holy Communion, or the Lord's Supper.* As Rheticus informed Eber, however, an even more serious theological controversy was brewing "about the divinity and humanity of Christ."

In the theological melting pot of Kraków in the early 1560s, some elements within Protestantism were actually veering off into the dangerous antitrinitarian heresy of Arianism, or Socinianism as it would later be called (after the Pole Faustus Socinus, 1539–1604). This teaching emphasized Christ's humanity at the expense of his divinity and denied his equality with God the Father. Although Rheticus was obviously pleased to have news to share with Eber, his theologically learned friend, his letter offered no details regarding the controversy and in fact expressed a lack of interest in them. Rheticus wanted, as he

---

*The nature of the sacrament was one of the matters of dispute between Calvinists and Lutherans. Eber tried to patch over some of the differences—and for his trouble was eventually accused of being a closet Calvinist and, in the late 1560s, in the years before his death, excluded from communion with the Lutherans.

put it, to be "just a Christian" and to get on with his work undistracted by sectarian battles. "Why cannot all intelligent people merely strive to live aright, and to do what is pleasing to God and useful to their fellow men?"

Rheticus's spirits fluctuated sharply during the early 1560s. In the letter of November 1, 1562, Rheticus had said how he longed for a friend. But only eighteen days later he cheerfully wrote Eber again to tell him about a Johannes Reinisch, who once attended Eber's lectures: "a good man, devoted to Luther and the Wittenbergers, and my friend."[7] This was barely two months before Rheticus's letter to Joachim Camerarius Jr. complaining about a dearth of books and grumbling about having to "sit here in the cave of the Cyclops."

Given his mixed feelings about Kraków, it could be asked why Rheticus stayed for so long in a place he often held in such low esteem. He did have other options: He could have gone to Vienna in 1553 or 1554, and even after he thought he was done with academia, his publications kept his name in wide circulation among educated people across Europe. Almost ten years after declining the call to Vienna, he wrote yet another letter to Eber (August 20, 1563), and on the envelope he scratched a postscript in German: "I have been invited to take a position in Rumania at four hundred talers a year plus full living expenses for six people for six years. But I won't go."[8]

Rheticus may have been happy enough in Kraków after all. Rumania would certainly have taken him farther away from the German heartland. In the end, it seems, Rheticus's frustration had two sources that had nothing to do with Kraków, a cosmopolitan city (both then and now) of immense charm. His principal source of dissatisfaction was simply the hurt of being in exile from his homeland. The other was his frustration at not making more tangible progress on his major work, especially the "Science of Triangles"—his chief bequest as the sole disciple and heir of Copernicus. He somehow could not focus adequately. It would seem that Fabricius had been right about mathematics and medicine.

Rheticus's Copernican motivation appeared vividly in another letter he wrote in 1563 to a man named Thaddeus Hájek (also Hagecius, 1525–1600), whom he had met in Prague and with whom he shared an interest in astrology. "I have again picked up the work of Copernicus," he reported, "and am considering elucidating it with a commentary . . . Some friends are asking and urging me to take up this task. If you can contribute any work to this project, please do so." The letter reemphasized Rheticus's persistent desire to bequeath something to posterity.[9] Yet the more strongly he felt this desire, the more frustrated he grew at seeing his projects languishing incomplete.

Rheticus's frustration with regard to those projects often mingled with vague longing and remorse. As Rheticus piously wrote in a letter to Eber on April 12, 1564: For the present, "by God's grace things are going quite well for me here, though . . . besides God alone, I have no one else who is supporting my work. May he deal with me as a merciful father, as he sees fit. Please remember me in your prayers and in those of your church."[10] As for the past: "We can censure [it] but not amend it. If we could, then I should be growing old in Wittenberg, where I spent my youth." The real piece of news in this letter, however, concerned the uncertain future. Rheticus closed by telling Eber: "France is calling me. But I have not yet decided what I shall do. Farewell."

Rheticus's "call" from France revealed just how widely Rheticus's work was attracting curiosity and admiration from the great centers of Europe. Leading citizens of what humanists called the "republic of letters," which encompassed sciences like geometry and astronomy, longed for Rheticus to fulfill his Copernican promise. Significantly, the Parisian episode began with Rheticus's groundbreaking publication on trigonometry: the *Canon of the Science of Triangles* (1551). This work had just reappeared, or was about to reappear, from the Basel printing house of Heinrich Petri.*

Apart from Rheticus himself, the main characters in the Parisian episode were Petrus Ramus (1515–72) and, in a supporting role, Jacques Caloni Portanus. Caloni had studied with Rheticus in Kraków on the advice of Johannes Crato. From Kraków he had moved on to Vienna, where, by his own admission, he accomplished nothing, and then returned to France. In Paris Caloni met Ramus, professor of eloquence and philosophy, and sang for him the praises of his former teacher. This meeting generated the idea that Rheticus might be an ideal candidate for the regius chair in mathematics at the University of Paris. On August 17, 1563, Caloni, with Ramus's blessing, sent to Rheticus from Paris a letter in which he reminisced about his time in Kraków, mentioned the possible professorship, and told Rheticus—in four affectionate words that summed up the whole letter—"wish you were here."[11]

As for Ramus, he himself was one of the most daring innovators of his generation in the fields of logic and rhetoric. His influence extended across European thought in his own and the following century. Most famously, he attacked the entrenched teachings of Aristotle, who was still routinely referred to simply as "the philosopher." The thesis Ramus defended for his M.A. at the University of Paris in 1536 (and in those days it really was a thesis, that is, a single proposition) was *Quaecumque ab Aristotele dicta sunt, commentitia sunt*, or, "Everything that Aristotle said is false."

Ramus's examiners, however, who were all dyed-in-the-wool Aristotelians, had to give him the degree. For in Aristotelian logic, it is not

---

*This second edition of the *Canon* was undated. Karl Heinz Burmeister estimates its appearance at 1565; see *Georg Joachim Rheticus, 1514–74: Eine Bio-Bibliographie* (Wiesbaden: Guido Pressler, 1967–68), 2:78. A year later the same house would publish the second edition of Copernicus's *Revolutions,* along with the third edition of Rheticus's *First Account.*

permissible to support or oppose an argument by presuming as certain the very issue whose truth or falsehood is at stake in the argument. That would be an instance of circular reasoning, the fallacy of "begging the question." What was at stake in Ramus's thesis was everything that Aristotle said. The hapless Aristotelian professors were therefore prevented by their own principles from adducing any Aristotelian evidence in opposition to Ramus's categorical thesis, and the twenty-one-year-old iconoclast was granted his M.A.

In 1563, two and a half decades later, Ramus saw Rheticus as representing hope for a new dispensation in another of the liberal arts: astronomy. On August 25, barely a week after Caloni's note floating the Paris scenario, Ramus wrote a long letter to Rheticus outlining his fuller vision.[12] Ramus had longed for the reformation of the disciplines, including geometry and astronomy. But in astronomy, up to this point, he could scarcely see the way forward. Despite his desire for progress, Ramus explained, he received "adequate help from neither logic, nor books, nor people. All [he] could see was a discipline whose complexities and obscure hypotheses rendered it incomprehensible." But his discovery of Rheticus's *Canon* changed all that. Ramus read and reread it, and, "wonderfully, it awoke in [him] a great hope" for an astronomy liberated from previous incumbrances.

The letter went on to tell how young Caloni, upon returning from his travels, provided the eager Ramus with a much fuller picture of Rheticus's "unique erudition" and of his writings on triangles and astronomy. Caloni's account left Ramus "inflamed with desire" for Rheticus. And hearing that Rheticus might consider visiting France, Ramus was moved to pen this letter: "to declare my love for you and to promise the most thorough, affectionate hospitality, should you come to Paris," and also "to exhort and encourage you not to let perish the extraordinary contributions of your mind, so valuable for all future generations, but to share them with the world at once, so that your fame should increase even more."

ITEM, DE LIBRIS REVOLVTIONVM NICOLAI
Copernici Narratio prima, per M. Georgium Ioachi-
mum Rheticum ad D. Ioan. Schone-
rum scripta.

Detail from the title page of The Revolutions *(1566).*

Ramus laid out grand prospects of fame and accomplishment, portray-
ing Rheticus as heroic liberator of astronomy. If Rheticus could solve its
fundamental problems—like a man "either untangling some Gordian
knot or else cutting it asunder"—he would be acknowledged, as the
Parisian professor of eloquence declared, "supreme ruler of Astronomy."

Despite its rhetorical flourishes, Ramus's nomination of Rheticus as a
kind of astronomical prince was solidly credible. It was Rheticus, after all,
who had introduced Copernican astronomy to the world of learning, initially by publishing the *First Account* (1540 and 1541), and soon thereafter by serving as what Giese had called chief impresario of *The
Revolutions* (1543). Within three years of Ramus's 1563 letter, Rheticus's
name and work were again publicly associated with Copernicus. For in
1566, in Basel, the second edition of *The Revolutions* appeared, and a further edition of Rheticus's own *First Account* was included as part of this
publication.[13] With regard to Copernican astronomy, Rheticus was finally emerging from behind the scenes. His contribution was announced
at the center of the title page of the Basel *Revolutions*: "Also, concerning
the Books of the Revolutions of Nicolaus Copernicus, a First Account
written by Master Georg Joachim Rheticus to Mr. Johann Schöner."

As for the call to astronomical heroism—and to Paris—in 1563,
Rheticus could hardly have remained unmoved by Ramus's passionate
appeal. Yet it did not move him from Kraków.* It did, however, evoke

---

*It was fortunate for Rheticus that he declined Ramus's invitation. Both men were Protestants, and in Au-
gust of 1572 Ramus was one of the thousands of Protestants, known in France as Huguenots, murdered in
the St. Bartholomew's Day Massacre.

from him a manifesto concerning exactly how he intended to perform the scientific exploits to which the great Parisian had called him. Not only did Rheticus write to Ramus with his declaration of intent; he gave a copy to a fellow Krakóvian, Jan Lasicki, who carried it all the way to Heidelberg. From there, on August 8, 1568, Lasicki wrote two letters, one to Heinrich Bullinger, the spiritual leader of Zurich, telling him he intended to send Josiah Simler, literary executor to Conrad Gesner (who had died in 1565), "a catalogue of the writings of the most outstanding mathematician of his generation, Joachim Rheticus . . . so that he might insert it into Gesner's collection." In the other letter, sent to Simler accompanying Rheticus's manifesto, Lasicki stated, "[The document] was handed over to me . . . by the foremost mathematician of our times, Joachim Rheticus, who lives as I do in Kraków." If it were included in the republication of Gesner's work, he added, it would "not so much gratify this Rheticus (when he finds out about it) as prove useful to readers."[14]

By divulging his goals to Ramus and releasing his letter into the public domain, not only was Rheticus further raising expectations for himself; he was taking the risk, should he not fulfill his promise, of appearing one of the most overly optimistic self-promoters of the sixteenth century.

Simler's introduction of the extract, published in his revision of Gesner's *Bibliotheca* in 1568, two years after the second edition of *The Revolutions*, underlined its author's status as the first Copernican: "Georg Joachim Rheticus, doctor and mathematician, offered the first account of the book concerning the revolutions by Nicolaus Copernicus of Toruń, whereby it is asserted that the earth moves about the heavens." And right after this Simler printed the manifesto, addressed to Ramus.[15]

Rheticus began this document by declaring his aim, as part of a fundamental reexamination of astronomy and geography, of offering a renovated science of triangles. "I have worked out the roots of the canon [of triangles] by means of algebra," he asserted, while admitting that much remained to be done toward establishing values expressed in ratios over a million billion (i.e., to fifteen decimal places). The work so far, he stated, had taken a dozen years, "and that with computational workers continuously on my payroll." To these accomplishments Rheticus also wanted to offer "a new collection of chapters on the science of spherical triangles."

At the center of the manifesto Rheticus promised a two-step program for improving astronomy based on careful observation—precise determination of "the positions of the stars, lights, planets, comets, and everything else that can be observed on high"—and exact calculation based on his science of triangles. If commercial accountants took such pains in calculating mere business transactions, Rheticus asked, "why should not we, in pursuing higher things, likewise also devise powerful and precise means for making our calculations?"

The most shocking statements in Rheticus's letter to Ramus concerned Ptolemy. The *First Account*, reissued in 1566, had called Ptolemy "divine" and "the father of astronomy,"[16] yet by 1568 the same Ptolemy appeared as a failure, and Rheticus devoted the letter's most rhetorical passage to insulting him: "As the huts that children build out of mud and sand compare with the buildings of Vitruvius or the palaces of Rome in its glory, so the majestic edifice of Ptolemy . . . compares with the true and reliable science of stellar motion." Rheticus sharpened his mockery by playing on the title of Ptolemy's astronomical treatise, the *Almagest*, which he implied is not in fact majestic but instead "a magnificent ruin."

Near the end of his letter Rheticus indicated to Ramus that he had almost arrived at "a new method for conducting natural science," including medicine. "And since I take great delight in chemistry, I have dug down to the foundations of this science"—both the mention of chemistry and the metaphor of "digging down" being indications of his Paracelsian interests. Finally, all of these ambitious research projects,

Rheticus claimed, were supported financially by "the practice of medicine, my Maecenas."

For his part, Ramus continued to hold a superlative and respectful opinion of Rheticus. While he harbored no resentment at his decision not to "adorn the University of Paris," however, there was disappointment.

In 1569, a year after Rheticus's manifesto appeared in the newly expanded *Bibliotheca* of Gesner and Simler, Ramus himself published another book, *Schools of Mathematics*, which surveyed the greatest mathematicians and mathematical achievements of his day. The survey included Rheticus and mentioned once more Ramus's cherished "hope of liberating astronomy from hypotheses." But Ramus—like Fabricius before him, and even perhaps like Rheticus himself in his more realistic moments—expressed regret over the service demanded of Rheticus by the Maecenas, the patron, the taskmaster Medicine. Ramus left no doubt that he saw Rheticus's mathematical potential as unfulfilled, which indeed at that time it was.

Ramus, like more and more of his contemporaries, could see that Rheticus had a choice to make. He could no longer serve two masters, not if he were to realize his mathematical and astronomical promise. Somehow he had to shake free—or be shaken free—of medicine. As of 1569, that potentiality appeared thwarted. As Ramus lamented: Were it not for the demands of medicine upon the time and the mind of Rheticus, "mathematics would now for a long time be proclaiming another Copernicus."[17]

CHAPTER 15

RESCUING RHETICUS

Despite remarkable things already accomplished, by the late 1560s Rheticus's career stood at a crossroads. The appearance of his own mathematical manifesto in Gesner's *Bibliotheca* of 1568, and then Ramus's extraordinary reference to him in *Schools of Mathematics* in 1569, placed Rheticus squarely on the mathematical map of Europe. Even if Ramus was inclined to emphasize his unfulfilled potential, Rheticus's reputation as one of the greatest mathematicians of his generation extended from Paris to the Black Sea.

That reputation also encompassed the two main domains within which sixteenth-century mathematicians and astronomers functioned: the universities and the courts of power.[1] Before leaving academia proper, Rheticus had already sought patronage from Duke Albrecht of Prussia, as well as from Saxon nobility such as Georg von Komerstadt and Christoph von Carlowitz. Then in the 1550s the wealthy Johannes Boner had helped him erect his obelisk in Kraków. Beyond that, there was the Vienna connection, which included both the offer of an academic position and, after he had turned that down, his 1557 letter seeking the favor of King Ferdinand, who a year later became emperor. (For an excerpt of this letter, see the appendix, item 8.)

Petrus Ramus, as a university reformer, naturally emphasized Rheticus's academic pedigree. His 1569 *Schools of Mathematics* highlighted

the huge part Nuremberg had played in nurturing the legacy of Regiomontanus. Ramus lauded the role of Melanchthon in building up a
model institution (Wittenberg) "outstanding not only for theology and
eloquence, but also for the mathematical disciplines"—a model that
had "kindled" the study of mathematics in many universities across
Germany.[2] Rheticus was one of its products. And even though by this
time he was in exile in Poland, he was, in Ramus's words, "lending
glory to Kraków with his mathematics."

Whatever the setting, Rheticus's own contributions to mathematics were more than ever being threatened by his need to serve his
Maecenas—his supporting patron—the lucrative profession of medicine. Rheticus's engagement with medicine was much more than merely
professional and financial; it was also deeply intellectual. The medicine
that was absorbing the mind of Rheticus was not safe, traditional
Galenic practice. It was the novel and highly controversial medical theory of Paracelsus. Rheticus's activity as a doctor was therefore an even
greater threat to his pursuit of mathematics than the perceptive Ramus
realized.

Rheticus expressed his deep curiosity concerning Paracelsianism in a
letter to his friend Thaddeus Hájek in Prague in 1567. "I hope you are
a keen reader of the writings of Theophrastus Paracelsus," he wrote;
"for I too delight in his works, and I would love to discuss them with
you." Rheticus drew his friend's attention to an astronomical work by
Paracelsus, of which he had read a fragment, and proposed that whoever first gained access to a full copy should immediately send one to
the other.[3] Typically, whenever Rheticus was interested in a subject, he
sought not only to grasp it as comprehensively as he could, but also to
promote it among his friends and the wider community of learning.

He did this both by sharing books and by writing up and publishing
his own research. As he had stated in his manifesto to Ramus, "Since I
take great delight in chemistry, I have dug down to the foundations of
this science, with the result that I have sketched out seven chapters

concerning it." This work has not survived, but there is substantial evidence that Rheticus was actively involved in extending the literature of Paracelsianism in the late 1560s and early 1570s.[4]

Paracelsian medicine emphasized cures based chiefly on minerals rather than plants, and for this reason a Paracelsian researcher based in Kraków had an advantage given the proximity of the great salt mines at Wieliczka. In another lost work, *De salinis,* Rheticus extolled the usefulness of salts and minerals from the mines. A seventeenth-century biographer of Copernicus named Simon Starowolski would assert that these metallic salts from Wieliczka are "known throughout Europe, the more so on account of the description given by the renowned mathematician Rheticus."[5]

The cavernous mines of Wieleczka, however, could also function as a symbol of the more ominous aspects of Paracelsian medicine. Difficult as it may now be to imagine, many in the early modern age found the practice of mining profane, and even rather hellish. Rheticus's comment that he had "dug down to the foundations" of medical chemistry could be read as more than a mere figure of speech. Even a century later, the poet Milton would identify the fallen angel Mammon as the diabolical instigator of mining.

> *By him first*
> *Men also, and by his suggestion taught,*
> *Ransacked the center, and with impious hands*
> *Rifled the bowels of their mother Earth*
> *For treasures better hid.*

The potentially scandalous associations of Paracelsianism went beyond the indecencies involved in rifling the bowels of mother earth. They included a taint of heresy of the most serious kind, one that involved the sensitive theological issue centering on the nature of the Son of God. It was Rheticus's friend from Zurich Conrad Gesner who had first proposed this line of attack, alleging in a letter to Johannes Crato on August 16, 1561, that "Paracelsians . . . deny the divinity of Christ . . . They seek to convince people that Christ was utterly and merely human."[6]

Crato too soon echoed this refrain in 1563 in a new edition of his *Methods of Therapy*, a work whose first edition Rheticus had requested back in 1554. Crato boldly alleged that the Paracelsians were guilty of Arianism, the same heresy for which Michael Servetus had been burned at the stake in Geneva only ten years earlier: "Certain people are weaving an infernal medicine shrouded in illusion and empty verbiage, and smelling of alchemical and barbaric muck. Such people are not so much doctors as obfuscators . . . even feigning that the Son of God, Savior of the human race, is the spirit of the world and of our bodies, and trying to cloak their Arianism."[7]

Another leading anti-Paracelsian in the early 1560s was likewise an old friend of Rheticus: Achilles Gasser, who had meanwhile moved from Rhaetia to Augsburg. Together with a fellow Augsburg physician, Gasser wrote a series of "Antitheses" to the "Propositions" of an avid Paracelsian named Alexander von Suchten—to whom Gesner gleefully sent a copy of his and his colleague's work. In a letter written on April 28, 1564, Conrad Gesner complimented Gasser on his efforts, calling him "the best defender of Hippocrates and Galen."[8]

It was undoubtedly Crato, as the emperor's personal physician, who wielded the greatest anti-Paracelsian influence. It is not surprising that the emperor himself was no friend of the new medicine. According to the influential anti-Paracelsian writer Thomas Erastus, who in turn claimed to be citing Crato, Ferdinand "regarded Paracelsus as a most mendacious and impudent impostor who had always refused contact with approved scholars."[9]

Given Rheticus's consistent efforts to ingratiate himself with royalty—as in the 1557 dedication to Ferdinand—the stigma of medical and with it theological unorthodoxy was a dangerous prospect, and his reputation may indeed have been besmirched as a result of his Paracelsian interests. A Polish author named Łukasz Górnicki (1527–1603), who came from an area near Kraków, described a kind of witticism whereby one removes from or adds to a word a single letter or short syllable in order to twist its meaning. To illustrate, he mentions how "one might [do this] with the name of Dr. Rheticus, thus calling him Hereticus."[10] In the last half of the sixteenth century such a witticism was potentially no laughing matter. Nothing indicates whether Górnicki himself simply invented the Rheticus/Hereticus play on words for his own purposes, or whether it was already a verbal burr that had attached itself to the name Rheticus.*

Part of the "outsider" profile of Paracelsus himself, though less serious than the taint of heresy, was the fact that he wrote not in Latin but in German. As a practical result, if his teachings were to be disseminated throughout Europe, his works must be translated from the vernacular into the universal language of scholarship, which was Latin. Here too Rheticus's drive not only to understand but also to promote Paracelsianism played a role. In the early 1570s he devoted time to preparing Latin translations of two works, one by Paracelsus and another perhaps erroneously attributed to him, which bore the rather cryptic titles (respectively) *Archidoxa* ("Secrets") and *Liber vexationis* ("Book of Vexation").†

---

*In the sixteenth century the concept of heresy was also often linked to homosexuality. The German word for heresy, *Ketzerei*, was "the vernacular term most frequently used to describe same-sex sexual behavior in civic court records"; Helmut Puff, *Sodomy in Reformation Germany and Switzerland, 1400–1600* (Chicago: University of Chicago Press, 2003), p. 42.

†*Vexation* was a technical term relating to the alchemical transformation of minerals. Both of the works mentioned have been lost, with the intriguing exception of a fragment of the second one, a single leaf bound in a manuscript volume held by the National Library in Florence, Italy, and headed "Philippi Theophrasti Paracelsi. De Alchimia liber vexationis latine conscriptus per Georgium Joachimum Rhoeticum." The handwriting in this document is that of Rheticus himself. Just as interesting, on the manuscript leaf are marginal comments in Italian, likewise in Rheticus's hand—proof positive of his facility in his mother's language. See Karl Heinz Burmeister, *Georg Joachim Rhetikus, 1514–1574: Eine Bio-Bibliographie* (Wiesbaden: Guido Pressler, 1967–68), 2:23, and Burmeister, "Die chemischen Schriften des Georg Joachim Rhetikus," *Organon* 10 (1974): 180–181.

Ironically, then, within only a few years of laying out his grand program of mathematical and astronomical research, Rheticus was not only a busy Paracelsian doctor but also an active Paracelsian author. Those who still treasured his potential and long-awaited contribution to the mathematical sciences had reason for becoming increasingly distressed.

The richest evidence of such concern is found in the collected letters of the Hungarian nobleman András Dudith (1533–89), also known in Latin as Dudithius. Dudith resided for many years in Kraków as a diplomat in the service of Maximilian II, who had become emperor when Ferdinand died in 1564. He corresponded prolifically with learned men across Europe on topics diplomatic, religious, scientific, and personal. His keen interest in astronomy, as well as membership in the humanist élite of Kraków, put him in direct and regular contact with Rheticus.

Relations, however, between Rheticus and Dudith had turned sour precisely because of Rheticus's Paracelsianism, which Dudith found repugnant.

On March 29, 1569, Joachim Camerarius Sr., who had spent part of the previous year in Vienna advising the emperor on religious affairs, sent a letter from Leipzig to Kraków in which he expressed puzzlement that Dudith, so interested in astronomy, nonetheless seemed not to be drawing as he might on the available expertise of Rheticus. Camerarius continued: "I have no doubt that your Excellency esteems Rheticus highly and converses freely with him as an outstanding expert in those things which you love so much. His prognostication, concerning which he wrote to me in Vienna, is so far showing no sign of completion—in fact the business rather seems to be going backwards."[11] Significantly, though acknowledging Rheticus's preeminent ability in astronomy, Camerarius betrayed disappointment at Rheticus's failure to complete projects related to astronomy.

In his May 15 reply the sophisticated Dudith described the condition of Rheticus in deliberately mythical language that called up all the infernal and heretical associations of Paracelsianism. "Do not be surprised that I do not use his services," he wrote to Camerarius. "For he is held fast by Theophrastus Paracelsus, so much so that he has now slid down from heaven into a manner of thinking whereby he apparently aims to strive with heaven itself and with nature—yes, with the very gods—thinking that Pluto can be called up from the underworld by the art of Vulcan. But I certainly find all steps in that direction quite terrifying."[12] Dudith wanted to stay clear of the impiety of a man who, despite his previous heavenly associations, had now, in his view, descended to a lower realm.

Dudith was not merely dismissing Rheticus out of hand, however. Despite his aversion to Rheticus's Paracelsianism, which he continued to describe in florid mythological language, Dudith was concerned about the potential scientific loss that would occur unless Rheticus returned to his true calling. Nor was Dudith's the only voice imploring Rheticus to fulfill that promise. As he wrote to Camerarius on February 8, 1570: "Rheticus does not cease to be an Argonaut, with the Swiss Theophrastus as his helmsman: and he keeps sailing onto the rocks . . . This grieves me, and I often natter at him, 'Let each one practice the art that he knows'—but it's no use. Thus neither [Johannes] Praetorius nor [Wolfgang] Schuler nor I have any success in calling him back to mathematics. We praise medicine and know that it is profitable, but in our view Theophrastus is not the place to investigate it."[13]

In 1571 Caspar Peucer, a mathematics professor and former student of Rheticus, publicly cited him as an example of how mathematical progress was being sadly impeded by mathematicians' lack of resources and by their involvement in other pursuits such as medicine. Peucer also drew explicit attention to Rheticus's in-progress "Science of Triangles" and its relevance to the astronomical legacy of Copernicus. "I am exhorting and imploring [Rheticus]," wrote Peucer, "to publish . . . the books he has composed concerning the science of triangles. For nobody has understood the mind of Copernicus better than he has, having sojourned with him." This was Rheticus's proper legacy, one that Peucer longed he should fulfill: "May he," his former student wrote respectfully, "by restoring the science of astronomy honor and immortalize his name to coming generations."[14]

Later the same year another acquaintance of Rheticus hinted in a letter to Dudith at how Rheticus's distraction by Paracelsian medicine was threatening in a specific way to hold back scientific progress. The Strasbourg mathematician and astronomer Conrad Dasypodius (1531–1601), who also corresponded with Petrus Ramus and with members of the imperial court in Vienna, had exchanged letters with Rheticus. Whatever was said in that lost correspondence had caused Dasypodius likewise to desire the completion of Rheticus's mathematical work.

Dasypodius was no mere curious onlooker; he had just been commissioned to redesign the astronomical clock in the Strasbourg cathedral,[15] a practical business that, along with Dasypodius's other mathematical projects, might well benefit from the latest trigonometric and astronomical research—assuming this were available.[16] But with Rheticus being swallowed up by Paracelsus, it was not. So on December 3 Dasypodius wrote in desperation to Dudith, "I urgently beg you to admonish the most learned Mr. Joachim Rheticus concerning his duty and the promise he made—that he would publish that about which he has often written to me."[17] Dasypodius did not know that Dudith's influence with Rheticus had dwindled to nothing. Two little Latin words from Dudith's correspondence with Camerarius summed up all his efforts, and all the efforts of others so far, to dislodge Rheticus from his growing Paracelsian obsession and refocus his efforts on his grand calling: *sed frustra*—"but in vain." And Rheticus, who was making plans to leave Kraków, would soon be even farther from the voices calling him to complete his Copernican mission.

* * *

The most resounding singular "voice" in all of sixteenth-century astronomy was heard for the first time on November 6, 1572. Its message was emitted from ten thousand light-years away, although this distance was neither known nor calculable at the time. That evening a little-known astronomer at the University of Wittenberg named Wolfgang Schuler was observing the constellation Cassiopeia and noticed a new star.

Schuler had resided for a time in Kraków, where he knew Rheticus. It was this same Schuler, according to Dudith's letter to Camerarius in February of 1570, who along with Johannes Praetorius and Dudith himself had been trying to "call [Rheticus] back to mathematics." Both Schuler and Praetorius in 1571 had themselves been called back to Wittenberg, where they now held professorships of mathematics.

Five days after Schuler first saw the new star, a twenty-six-year-old Danish astronomer named Tycho Brahe also noticed it. So thorough and perceptive was the Dane's account of the phenomenon—which made him as well as the star famous—that it is still known as Tycho's Supernova. After his initial expressions of disbelief, Tycho went on to offer a scientific account of his observations conducted over an eighteen-month period. The stunning message of his study was that the claim, as he put it, about which "all philosophers agree"—"that in the ethereal region of the celestial world no change . . . takes place"—was false.[18] One of the main assumptions undergirding two thousand years of astronomy was crumbling, and the new star called more clearly than any human voice could for a new astronomy.

It would also strengthen the call for a new, more powerful system of angular measurement and calculation. Tycho, four decades before the invention of the telescope, measured the position of the nova to an accuracy of one twentieth of a degree—and found that the star's position did not change. A finer scale and more precise mathematical tools were required if a more convincing new picture of the universe of stars were to be produced.

The appearance of the new star caused a sensation among astronomers all across Europe, from John Dee in England to Philip Apian and Michael Maestlin in Tübingen and Thaddeus Hájek in Prague (correspondent of Rheticus, Dudith, and Tycho Brahe). It also occasioned vigorous discussion in court circles, such as those of Wilhelm IV, landgrave of Hesse, and Ludwig, duke of Württemberg.[19] The likelihood of a well-connected astronomer like Rheticus hearing nothing about it—even beyond the High Tatra Mountains—would have been very low. It was even less likely that a mathematician or astronomer living right in Wittenberg would have failed to hear the news. In this small Saxon town the nova was commented on by not only Schuler but also Rheticus's former student Peucer.* Another resident of Wittenberg in 1572, a student of Schuler, Praetorius, and Peucer, was a promising young scientist named Valentin Otto.

For reasons that are not fully understood, Rheticus left Kraków in 1572 and traveled to the region known at Scepusia, southeast of the High Tatra Mountains, to the Upper Hungarian cathedral town of Cassovia.† He had been offered the patronage of Baron Johann (or János) Rueber, a German-speaking Protestant closely tied to the imperial court in Vienna. Emperor Maximilian II himself referred to Rueber as "our captain general in the upper parts of the Kingdom of Hungary."[20]

Rheticus had honest intentions of carrying on with his "Science of Triangles" in Cassovia, for he took many (though not all) of his relevant documents along with him when he left Kraków. Yet Paracelsian medical theory continued to absorb his attention and his time. As Dudith wrote to Thaddeus Hájek in Prague on April 12, 1573, "Rheticus is still away in Hungary and an admirer of Paracelsus."[21]

---

*In 1566 Peucer had been Tycho Brahe's first astronomy teacher during the Dane's brief time as a Wittenberg undergraduate.
†Today Cassovia is known as Košice and is the principal city of eastern Slovakia.

Early seventeenth-century depiction of Cassovia: "the principal city of Upper Hungary."

At about the same time, in Wittenberg, Valentin Otto was nurturing something few people ever experience: an intense passion for geometry. Despite his youth, not only had Otto mastered geometry as it was then taught; he was already searching for ways to extend the field beyond its existing boundaries. "In spite of how much Euclid teaches us about triangles," Otto wrote, "his *Elements* offer us no way of deriving the angles from the sides, or conversely the sides from the angles."*

One of Otto's teachers, Johannes Praetorius, who in 1569 had spent some time in Kraków in close contact with Rheticus, knew about his monumental but unfinished "Science of Triangles."[22] Otto's most direct introduction to Rheticus, however, came when he discovered on his own the 1551 pamphlet *Canon of the Science of Triangles*.

As Otto recounted: "[I realized that] unless we better understood the science of triangles, the way forward was blocked. I therefore began

---

*This deficiency can still be appreciated by anyone who has ever tried to solve problems in high school trigonometry without the aid of a calculator or trig tables.

to ponder how I might best satisfy this need . . . But with great luck I stumbled upon the dialogue of Rheticus which he attached to his *Canon*. Reading and re-reading it, I sensed with wonder that it offered lessons—many indeed—truly worth learning. Aroused and inflamed by these things, I could not restrain myself, but had to seize the first chance I could to visit their author and make his acquaintance personally. So I set out for Hungary, where Rheticus was still staying."[23]

Rheticus's response to the arrival of Otto on his doorstep in Cassovia was more than the enthusiastic young pilgrim from Wittenberg could have anticipated. He "received me most warmly," Otto wrote. "But before more than a few sentences had been exchanged between us, he grasped my reason for coming. He exclaimed: 'You are the same age, coming to visit me, as I was when I visited Copernicus! Were it not for my journey, his work would never have seen the light of day.' "

Although there is little reason to doubt the accuracy of Otto's account of his arrival in Cassovia, it is clearly selective and incomplete. Otto—with some justification—structured his narrative around how his relationship to Rheticus paralleled Rheticus's relationship to Copernicus. Crucially, that parallel served as a sudden revelation that prompted Rheticus to resume serious work on triangles when all other efforts to do so had failed.

Otto's account, however, while movingly and poetically symmetrical, remained almost silent concerning the interwoven social and scientific conditions that enabled both Rheticus and Otto to rescue the careers of their adopted teachers. Both men undertook heroic individual missions, yet those missions sprang up out of an extensive, rich undergrowth of scientific activity and curiosity.

In the case of Rheticus, some details are known about the motivations—and motivators—that spurred his visit to Copernicus in 1539, such as his contacts over the previous months with Schöner, Petreius, and Gasser. Concerning the background of Otto's visit to

Rheticus thirty-five years later, nothing is certain apart from the story of Otto's excitement at reading the *Canon*. Circumstantial evidence, however, suggests the possibility that something like a "benevolent scientific conspiracy" lay behind the mission to rescue Rheticus.

The warrantable circumstantial evidence is this: Rheticus had made his trigonometric ambitions completely clear by publishing the *Canon of the Science of Triangles* in 1551, and then by writing his mathematical manifesto to Ramus in 1568 and permitting its publication in Gesner and Simler's *Bibliotheca*. In a comment about the *Canon* that could also apply to the manifesto, Otto pointed for a moment beyond the leading individual actors and out toward the larger scientific scene: Rheticus's work, he wrote, "aroused wonder as well as both hope and expectation among the most learned men."[24]

These details do not by themselves clinch the case for a benevolent scientific "conspiracy" aimed at pressing Rheticus back into the service of mathematics. Yet despite Otto's almost complete silence about the role of the scientific community in enabling his visit to Cassovia, the unfolding of his relationship with Rheticus was highly consistent with the hypothesis that a network linking (at least) Wittenberg, Kraków, and Vienna helped to launch Otto on a mission to rescue Rheticus and his mathematical work—a mission whose imperial infrastructure was already in the process of being assembled.

Whatever sequence of events delivered Otto to Cassovia in that springtime of 1574, his appearance rekindled Rheticus's sense of mathematical mission as nothing else had been able to do. The igniting spark, of course, was Otto's youthful passion, in which Rheticus recognized a picture of his own enthusiasm thirty-five years earlier when he made his pilgrimage to Frauenburg to meet Copernicus. He recalled both the human dimension and the scientific consequences of that meeting: the powerful student-teacher bond he formed with the beloved astronomer, and the completion and publication of Copernicus's masterpiece.

Rheticus's reception of Otto also made good on the promise implicit in his 1551 *Canon of the Science of Triangles*. At the end of its dialogue, Philomathes had assured the mathematically curious Hospes that he could expect a warm welcome from Rheticus, for "nothing pleases him more than those who truly cherish [mathematics]" and "desire to promulgate its sound principles." In Wittenberg Otto must have recognized himself in this description; in Cassovia—consciously or not—Rheticus was fulfilling the destiny he had scripted for himself twenty-three years earlier.

As for Otto, he soon proved to be a useful Philomathes as well as a symbolic Hospes. The "Science of Triangles" had demanded much labor and skill, and it would continue to do so. Part of what had held it back was the expense and difficulty Rheticus repeatedly encountered of finding the right help—and people with an adequate mathematical education. But Valentin Otto offered a gratifying contrast to this pattern. As he would later recount, he and Rheticus got right down to work, dividing the labor between them. Rheticus continued his study of the Islamic mathematicians (especially the great al-Battani), while Otto pursued his apprenticeship by "reading and rereading the [sections from the 'Science of Triangles'] that were already finished."

Despite the brevity of Otto's account of these labors, they apparently consumed all of the summer and most of the autumn. By then, however, yet another journey was required, one that would prepare the way for the climax of Otto's narration. When Rheticus "had almost finished the task of working out the first and second series of the canon of the science of triangles," Otto wrote, "he required access to things he had left behind in Kraków. Since he would not consider entrusting these things to anyone else, the job of transporting them fell to me."

Otto then set off along a challenging and serpentine route to Kraków on the other side of the High Tatras. It proved to be much too late in the year for a reasonably safe transit of the mountains and valleys, and the dangers of the trip offered a foreshadowing of those that would soon befall Rheticus himself. Otto wrote: "During my journey, unexpected and continuous rain fell for several days and nights, which made travel not only difficult but also extremely perilous. Twice in one day I was in danger of drowning."

Otto survived his journey, and on November 28, 1574, he returned safely to Cassovia bearing his precious trigonometric cargo—only to find that in the meantime Rheticus had encountered perils of an even more serious kind. During Otto's absence Rheticus had been invited to stay at the residence of Baron Rueber. But there he had "slept in a recently plastered room" and, affected by the damp, contracted a respiratory infection, from which he was still suffering when his student reappeared.

A few days following Otto's return, Rheticus was again summoned by the baron. But by this time, amid continuing bad weather in Cassovia, Rheticus's condition had worsened. As Otto would recount: Rheticus, realizing his life was coming to an end, "sent his friends to Lord Rueber, requesting that once he died, his work . . . be left to me . . . so that I could finish it where and when I might. For he did not doubt but that I had the necessary resources to do the job and to deliver the completed and finished work to posterity as soon as possible, especially since I had given him my sacred promise." Rueber consented.

Finally, Otto wrote, "after four more days, the illness growing worse and worse, at about two o'clock during the night, the most beloved teacher Georg Joachim Rheticus expired in my arms." Rheticus, the first Copernican, famous astronomer and mathematician, died on Saturday, December 4, 1574, two months short of his sixty-first birthday. He had prepared himself for death and had kept a clear mind. In his last days he was comforted by someone who loved him and would honor his dying wishes. Just in time, the rescue of his life's work had successfully begun.

CHAPTER 16

TRIANGLES, STARS, AND THE
SWEETNESS OF THINGS

At the end of 1574 Rheticus was dead, and his "Science of Triangles"—
a project aimed at enabling the exact calculations needed for "a precise
account of the starry sphere"[1]—lay unfinished and unpublished in a
remote region of Upper Hungary.

The body of Rheticus was buried somewhere in or near Cassovia; no
exact location was recorded. The files and papers that embodied his
work on triangles, however—along with other significant documents—
were safely and legally in the possession of Valentin Otto. Whatever financial difficulties Rheticus had faced in pursuing his work seemed to
have been swept away, if only temporarily.

Baron Rueber, captain general of Upper Hungary, had extraordinary confidence in the value of the work Rheticus had begun, and he
presented the matter directly to the Holy Roman Emperor himself.
Maximilian not only officially ratified the bequest of Rheticus, but also,
as Otto recounted, "beyond my every hope and expectation ordered
that I be provided with the funds required for finishing this work."[2]
While still grieving the loss of his beloved teacher, Otto found himself
legal proprietor of Rheticus's entire literary estate and recipient of a
royal grant with which to continue his program of research.

Like Rheticus, however, Otto encountered delays. Less than two
years after the imperial approval of his work, news came from Vienna

that Emperor Maximilian had died—and soon thereafter Otto's grant likewise came to an end. For a time Baron Rueber continued to support Otto from his own resources, though this arrangement was not sustainable. Before long—aided, no doubt, by the baron's excellent connections—Otto was called to a professorship of mathematics at the University of Wittenberg, where Prince-Elector Augustus of Saxony took up the sponsorship of Otto's project.

Otto's sojourn there was short-lived, for Wittenberg was torn by the theological controversy surrounding "Crypto-Calvinism," also known as "Philippism," so called after Melanchthon's Christian name. The reformer, who had died in 1560, was suspected of having fallen away from core Lutheran teaching concerning Holy Communion, and the cloud of alleged heresy continued to overshadow his followers; his son-in-law Peucer had already been imprisoned since 1576. Shortly after returning to Wittenberg, some time around 1580, Otto along with some other fellow professors found it (in his own understated words) "necessary to leave."

Otto then "consumed several years in wandering" but finally, on the advice of Peucer, who had just been released from ten years in prison, made his way to the Palatinate, where in May 1586 he was enrolled as a member of the University of Heidelberg and the following year became a professor. In Heidelberg, though impeded by periods of ill health, he continued working to complete Rheticus's trigonometrical project.

The culmination of these efforts was eagerly awaited by scholars and scientists across Europe. In 1593 the prominent mathematician Adrianus Romanus (Adriaan van Roomen) beat the drums for that "most highly desired" work, which Otto, he said, had promised him would appear "this year."[3] Romanus addressed these remarks to the Vatican astronomer Christoph Clavius and so helped keep Rheticus's name in currency on the mathematical front lines of Europe—still, in part, as a result of his 1568 manifesto to Ramus, which Romanus quoted enthusiastically and in full as evidence of the Herculean scope of the "Science of Triangles."

Otto, however, was proving to be a rather slow Hercules. The challenges of producing a work comprising almost fifteen hundred folio pages, a majority of these in tabular form, would have been truly daunting. It required the computation by hand of roughly a hundred thousand ratios to at least ten decimal places—or, more precisely, since decimals were not yet in use, ratios expressed as a number divided by ten billion. These functions were expressed for every 360th of a degree (that is, for every degree, minute, and interval of ten seconds) up to ninety degrees. During twelve of his Kraków years, Rheticus had employed five computers—not machines but people—merely to perform calculations. For Otto, organizing and keeping track of all these results, then typesetting them in orderly pages and columns—to say nothing of the labors of proofreading—was indeed a mammoth undertaking.

Finally, in 1596, the great work appeared under the title *Opus palatinum de triangulis,* in honor of Otto's most recent source of support, Prince Elector Frederick IV in Heidelberg, ruler of the Palatinate. Like many of Rheticus's publications, the title page of the *Opus* graphically highlights the function of obelisks in teaching (as Rheticus put it) "God's geometry in heaven and on earth." The astronomical significance of the *Opus* is made manifest by the presence of the sun, moon, and stars at the top of the title page, and at the bottom by pictures of instruments used for angular astronomical measurement: the quadrant and the Jacob's staff. From the top left corner, the sun emits a beam of light tracing a diagonal line that touches the pinnacle of the left obelisk and then disappears behind the title tableau. The extension of the line to the base of the obelisk on the right—if one imagines a horizontal baseline—would demarcate a 3-4-5 triangle.

The opening of Otto's preface likewise emphasized that the science of triangles would serve astronomy. "For by means of trial and practice,

Title page of Otto and Rheticus's Opus palatinum de triangulis *(1596).*

distinguished mathematicians ancient as well as more recent have learned that this one science opens the way to the noblest branch of philosophy concerning the motions of the heavenly bodies. Nor can anyone not steeped in the theory of the science of triangles penetrate to the secrets and mysteries of that divine art."*

The problem was that, once the *Opus* appeared and mathematicians had a chance to examine it in detail, disappointment began to be expressed, particularly concerning unreliability in its tables of tangents and secants. In July of 1598 Romanus wrote, "The *Opus Palatinum* . . . swarms with errors and (I say it frankly) is very misleading."[4] Such criticism spread in mathematical circles during the final years of the century, eventually reaching as far as Jan Broscius (Brożek) in Kraków. Of course it also reached Otto in Heidelberg and—embarrassingly— the man to whom he had dedicated the work, Frederick IV.

Not only was the *Opus* marred by calculation errors; its prose sections were breathtakingly long-winded and opaque.[5] A revision was called for, but sadly Otto was no longer up to the task. At the age of fifty, he was showing signs of physical and mental decline.[†] Although he did manage to find the energy for a return trip to Prague in 1600,[6] shortly thereafter, with the need for a new edition of the *Opus* impossible to ignore, he had withered to a shadow of his former vigorous, intelligent self.

---

*In a similar vein, the Englishman Raphe Handson reasserted a hundred years after the birth of Rheticus that trigonometry is not only rich as an object of "contemplation" but also still more "profitable in the practice." "For thereby all heights, depths, distances, questions of the map, globe, sphere, or astrolabe, may be more truly supputated [=calculated] than by any instrument whatsoever, besides the infinite use thereof in geometry, astronomy, [and] cosmography"; *Trigonometry, or, The Doctrine of Triangles First Written . . . by Bartholmew Pitiscus*, translated by Raphe Handson (London, 1614), A2r-v. More recently, Stanley L. Jaki has asserted that "the science of trigonometry . . . was in a sense a precursor of telescopes. It brought far-away objects within the compass of measurement and first made it possible . . . to penetrate in a quantitative manner the far reaches of space"; *The Relevance of Physics* (Chicago: University of Chicago Press, 1966), p. 189.

†Whether through bad judgment or bad luck, Otto in 1596 had been robbed by his own famulus only months before the appearance of the *Opus*. See Karl Heinz Burmeister, *Georg Joachim Rheticus, 1514–1574: Eine Bio-Bibliographie* (Wiesbaden: Guido Pressler, 1967–68), 1:180.

Elector Frederick turned instead to his own chaplain, a Calvinist theologian and accomplished mathematician by the name of Bartholomew Pitiscus, who had already composed his own book on triangles, titled *Trigonometry* (1595). It was this title that established the term by which the science of triangles would become universally known.

Pitiscus, like Romanus, realized that to get results accurate to ten places for functions of angles approaching zero or approaching ninety degrees, one must actually do the calculations to fifteen places. Somehow he learned from Otto that Rheticus had indeed prepared tables to fifteen places—yet Otto had neglected to use them in preparing the *Opus palatinum*. In the words of Pitiscus: "Enfeebled by a senile memory, [Otto] could no longer tell me where they were. He thought he had left them behind in Wittenberg."[7] Pitiscus sent an envoy to Wittenberg to look for the missing documents, but he returned, "after no small outlay of money, empty-handed." The search seemed hopeless.

In 1602, in his early fifties, Otto died, whereupon what was left of Rheticus's literary estate passed into the hands of Otto's friend Jacob Christmann, dean of arts at the University of Heidelberg. And "beyond all expectation," as Pitiscus would write, Christmann "discovered the very canon I was longing for. Therefore I too went to examine those papers that Rheticus had left behind. I excavated them one page at a time from their state of neglect, filthy and almost putrid. Irksome as this work was, I have had no regrets. For from [these documents] I have gleaned many things that have delighted me wonderfully."

Following this delightful find, Pitiscus devoted some further years to a revision of the *Opus palatinum*. The result was another monumental folio volume, this one titled *Mathematical Treasury: or, Canon of Sines for a Radius of 1,000,000,000,000,000 Units . . . as Formerly Computed at Incredible Effort and Cost by Georg Joachim Rheticus.*

Pitiscus thus successfully revised Rheticus's work with the help of Rheticus. As one modern historian of mathematics commented, these

CANON				SINUUM			
Sinus	Diff. I.	II.	III.	Sinus complementi	Diff. I.	II.	III.

(table of fifteen-decimal-place sine and complement values, Pitiscus 1613)

| Sinus complementi | Diff. I. | II. | III. | Sinus | Diff. I. | II. | III. |

76

A page from Bartholomew Pitiscus's Mathematical Treasury (1613), based on the
"canons" of Rheticus. The left column shows values to the equivalent of fifteen
decimal places for sines of angles at intervals of ten seconds (in this example, from
13° 0′ 0″ to 13° 9′ 50″).

"colossal" computations produced tables so usable that they were only superseded by works appearing early in the twentieth century.[8] A more recent historian of trigonometry states, Rheticus's tables "had astonishing accuracy—so much so that they were the basis of astronomical computation for centuries afterward . . . Rheticus systematized the process [of trigonometric calculation], turning it into a rigorous science rather than a piecemeal art."[9] Pitiscus's *Treasury*, embodying these accomplishments, was published in Frankfurt in 1613, ninety-nine years after the birth of the tables' original author.

Nor was that author's name entirely forgotten. In 1634 a book was published in Hamburg under the ambitious title *Trigonometric Key of the Universe*. Its author, Georg Ludwig Froben, paid homage to Rheticus by citing Rheticus himself right on the title page. The citation was borrowed from the *Canon of the Science of Triangles*, first published in 1551, and it remains a fitting summation not only of Rheticus's trigonometric legacy but also of the flowering of science during the century following the death of Copernicus: "To someone ignorant of geometry and arithmetic, [the numbers contained in these tables] are but mute ciphers. However, with anyone liberally educated in those disciplines they converse most sweetly. They teach him to behold with understanding the heavens and the earth and the natural objects they contain—so that everywhere he turns his eyes he knowingly grasps something regarding these excellent things, which consist by number and measure."[10]

The sweetness of mathematics and of the heavenly things that it reveals was a leitmotif of Rheticus's entire intellectual development: from his inaugural lecture in 1536, when he was twenty-two (arithmetic and geometry afford access to "the sweetness of things"), through his academic career (Melanchthon, 1542: Rheticus was "well suited to teaching those sweetest arts concerning the motions of the heavens"), and on into his long struggle to forge a science of triangles (Fabricius, 1554: "Serve mathematics—taste its sacred sweetness"). As indicated by Froben's use of his words, that sense of pleasure in geometry and in the harmonious numbers of astronomy remained part of Rheticus's legacy.

It was, moreover, a legacy that remained integral to the unfolding of the Copernican vision of the universe. In England in 1675, in a "Catalogue of the Most Eminent Astronomers, Ancient & Modern," Edward Sherburne described the career of Rheticus in the following way.

> GEORGIUS JOACHIMUS RHETICUS, Disciple to Copernicus, and Professor of Mathematicks in the University of Wittemberg . . . [From there he] went to Copernicus, whom he ceased not to exhort to perfect his Work, *De Revolutionibus*, which after his death he made publick, illustrating his Hypothesis by a particular narration, which he dedicated to Iohannes Schonerus, published by Maestlinus, and annexed to Kepler his *Mysterium Cosmographicum*, in the year 1621. He likewise set forth *Ephemerides*, according to the Doctrine of Copernicus, until the year 1551. What other Astronomical or Astrological Works he had either perfected or designed, will appear by his Epistle written to Petrus Ramus.[11]

Sherburne's entry points not only to Rheticus's direct links with Copernicus and Ramus, which were forged during his lifetime, but also to his posthumous association with Kepler, whose work was singularly vital to the further unfolding of Copernican theory.

In 1540 and 1541, when the first two editions of the *First Account* were published in Danzig and Basel respectively, Rheticus expected that once *The Revolutions* itself appeared, his own little book would fade into obsolescence. That, however, did not happen: Because it was shorter and livelier than Copernicus's masterpiece, and in some respects clearer, it continued to play an important role in educating people about the new cosmology. Its third edition coincided with publication of the second edition of *The Revolutions* (Basel, 1566), to which it was appended, minus the "Encomium on Prussia" but including the

Addita eſt erudita NARRATIO M. GEORGII IOACHIMI RHETICI, de Libris Reuolutionum, atqʒ admirandis de numero, ordine, & diſtantijs Sphærarum Mundi hypotheſibus, excellentiſſimi Mathematici, totiusqʒ Aſtronomiæ Reſtauratoriſ D. NICOLAI COPERNICI.

Detail from the title page of Johannes Kepler's Cosmographical Mystery *(1596): "To which is added the learned account of Master Georg Joachim Rheticus concerning the books of the revolutions, and concerning the wonderful hypotheses regarding the number, order, and distances of the spheres of the universe, according to that most excellent mathematician and restorer of all astronomy, Mr. Nicolaus Copernicus."*

prefatory letter by Gasser that had been added to the earlier Basel edition of Rheticus's primer.

This sensible situating of Rheticus's work with that of Copernicus did face one major obstacle. In 1559 the Roman *Index of Prohibited Books* had appeared, containing the names of Luther, Schöner, Rheticus, and just about every other known Protestant author. Copernicus's own work was not censored at all during the sixteenth century, but a dozen of the 324 extant copies of the 1566 *Revolutions* have had the *First Account* removed and in some cases Rheticus's name blotted out from the title page, where it originally appeared.[12]

Nevertheless, the *First Account* would be published twice as many times during the sixteenth century as *The Revolutions* itself. Its fourth edition appeared in 1596 bound together with a treatise that marked the beginning in earnest of the next wave of Copernicanism: Kepler's *Cosmographical Mystery*. This was the work famously proposing the outrageous idea that God had determined the dimensions of the spheres of what would now be called the planetary orbits in keeping with a nested series of the "Platonic solids"—regular geometrical shapes: cube, tetrahedron, dodecahedron, icosahedron, and octahedron.[13]

Unpromising as this thesis might seem, the link between Rheticus and Kepler as bold pioneers of Copernicanism was their shared commitment to geometry. The editor of Kepler's complete works suggests it was Rheticus's Copernican vision of the "celestial harmony" of spheres "geometrically defined"[14] that ignited the program of research to which Kepler devoted his life.[15] In the *First Account*, Rheticus had approvingly cited the pseudo-Platonic maxim "God ever geometrizes."[16] And in 1596 Michael Maestlin, Kepler's teacher in Tübingen who arranged for the copublication of that work with the *Cosmographical Mystery*, added a marginal note asking, "What would Rheticus have done had he noticed the geometry of God as regards the five regular solids that Kepler discusses?"[17]

Although this question is entirely hypothetical, no one can doubt the profound geometric contribution that Kepler eventually made to the unfolding and establishment of Copernican astronomy. Rheticus thus played a critical role not only in launching but also in extending the Scientific Revolution. In the words of one of the twentieth century's greatest expositors of Copernicanism, "The name of Joachim Rheticus runs like a golden thread through every stage of the early history of the impact of Copernicus."[18]

In the century following Rheticus's lifetime, as Sherburne's "Catalogue of the Most Eminent Astronomers" implied, the *First Account* appeared side by side with Kepler's work again, this time in the second edition of the *Cosmographical Mystery* (1621). Rheticus's Copernican primer—far from becoming obsolete with the appearance of *The Revolutions* in 1543—continued to play a vital role as a reader-friendly companion text to both first- and second-generation Copernican treatises.

The Revolutions embodied the work that Rheticus first traveled all the way to Frauenburg in 1539 to discover—and to urge Copernicus to complete. Rheticus then carried a transcript of it with him to Nuremberg, where the book was published, but Copernicus's original manuscript had stayed behind in Varmia. Upon the death of Copernicus, this treasure passed into the hands of his friend Tiedemann Giese, who after 1543 sent it on to Rheticus. When Rheticus died in 1574, it

Jacob Christmann's clear handwritten record of the provenance of Copernicus's manuscript of The Revolutions *(here called "de revolutionibus coelestibus").*

was inherited by Valentin Otto—and for years, Otto alone seems to have been aware that it survived as part of his teacher's estate.

Upon Otto's death in Heidelberg in 1602, Jacob Christmann excavated his books and papers, and there was the manuscript of *The Revolutions*. Christmann carefully recorded the work's provenance to that point: from Copernicus, to Rheticus, to Otto, to himself. As dean of the faculty of arts, he intended to make the manuscript available for the benefit of mathematical studies. Over the centuries, this singular document passed through the hands of a number of owners, including the Moravian educational reformer Jan Comenius (1592–1670). Finally in 1956 it was donated by the National Museum in Prague to the Jagiellonian Library in Kraków. Today it is one of three Polish artifacts designated under UNESCO's "Memory of the World" program, though no one might ever have heard of it, or even of Copernicus, had it not been for Georg Joachim Rheticus.

It was he, lover of stars and triangles, who became the first Copernican and delivered his teacher's priceless legacy to future generations.

ACKNOWLEDGMENTS

Research for this book was assisted by a grant from the Social Sciences and Humanities Research Council of Canada, for which I am most grateful. I also thank the Peter Wall Institute of Advanced Studies at the University of British Columbia for its support during my residence there in 2002. Much earlier, in 1990, I was awarded an Alexander von Humboldt fellowship for a year's study in Bonn, at which point the idea of writing this book had not entered my head. During that time in Germany, however, I learned to work with German sources and scholarship—resources without which a biography of Rheticus would have been impossible—and I thank the Humboldt Foundation for this gift. In addition, I would like to express deep gratitude to the two men whose friendship, with that of their families, has done more than anything else to cultivate my enjoyment of German culture and language: Götz Schmitz and Reiner Dienlin.

Many others have given valuable assistance as this project unfolded. For kind advice or encouragement I thank William Ashworth, Peter Barker, Len Bergren, Michael Crowe, Stefan Deschauer, Menso Folkerts, Stephen Guy-Bray, Kenneth Howell, Larry Molnar, Christoph Schöner, András Szabó, Katherine Tredwell, Glen van Brummelen, and my late colleague Stephen Straker. I also wish to acknowledge Robert Westman, whose article "The Melanchthon Circle, Rheticus, and the Wittenberg Interpretation of the Copernican Theory" in *Isis* in 1975 was a seminal contribution to the study of Rheticus in the English language. The other huge contribution that deserves recognition in this category is that of Edward Rosen, translator of

ACKNOWLEDGMENTS

(among other things) Rheticus's *First Account* (*Narratio Prima*); I would like to thank Mrs. Sandra Rosen for so graciously granting me free use of her late husband's translations and editions.

I am exceedingly grateful for the Polish hospitality I was shown in 2002 in Toruń, Warmia, and Kraków by Mira Bucholz, Edwin Świtała, Beata Tarnowska, and Grażyna Branny; for research assistance provided at the University of British Columbia by Rachel Poliquin and Grzegorz Danowski, as well as in Oxford by Alan Ward; and for generous assistance with translation by Stephanie Berlin, Tracy Deline, William Donahue, John Hale, Gernot Wieland, and Jens Zimmermann.

For assistance given in various libraries I also offer warm thanks: to the Bodleian (especially Duke Humphrey's Library and the Radcliffe Science Library) in Oxford; the Uppsala University Library in the Carolina Rediviva; the Österreichische Nationalbibliothek in Vienna; the Biblioteca Apostolica Vaticana in Rome; Biblioteka Jagiellońska in Kraków; the Universitätsbibliothek Leipzig; the Stadtbibliothek Feldkirch; the Stadtarchiv Lindau; the New York Public Library; and not least the Univeristy of British Columbia Library in Vancouver (especially the resourceful people in the Interlibrary Loans department). Thanks also to Bruce Bradley of the Linda Hall Library of Science, Engineering & Technology in Kansas City for help acquiring images, and to Kirsten Behee in Vancouver for her "scratchboard" rendering of Rheticus's (de Porris) coat of arms.

I owe a special debt of gratitude to Owen Gingerich, the foremost living scholar of Copernicus, who in recent years has been a faithful source of advice and encouragement. And I could never have written this book at all but for the groundbreaking archival and biographical work of Karl Heinz Burmeister—along with his generous personal provision of new material and insights as the project took shape. My greatest scholarly debt of gratitude is to him.

Further thanks are due to my agent, Peter Rubie, for persistence and skill as the book approached completion; to my incomparable editor, Michele Lee Amundsen; and to my extraordinary publisher, George Gibson. My friends Leo and Janet Pel were invaluable sources of encouragement—Leo in listening to my ramblings about Rheticus early every Sunday morning, and Janet in demanding that my prose be vivid and readable.

ACKNOWLEDGMENTS

Finally, for me, this book would also have been unimaginable without immense assistance given by members of my own family, including help with research from Jessa; with writing from Nora, Eva, and John; and with research, writing, sanity, and general well-being from the one who remains the music and the anchor of my life—to whom it is lovingly dedicated.

APPENDIX
TRANSLATIONS OF SOME DOCUMENTS RELATING TO THE CAREER OF RHETICUS

1. BISHOP TIEDEMANN GIESE'S NOTE TO DUKE ALBRECHT ACCOMPANYING A COPY OF THE *FIRST ACCOUNT*, APRIL 23, 1540

Since the astronomical speculations of the worthy gentleman Nicolaus Copernicus, canon of Frauenburg, because of their extraordinary novelty, seem strange to everyone, and have recently also stirred up a highly learned mathematician from the University of Wittenberg to come to this land of Prussia that he might investigate their foundation and cause—and now, in advance of the appearance of the new astronomy of that gentleman doctor, has published in the form of a small book a short account and preliminary announcement, in which he has not neglected also to praise this land and make glorious mention of the name of Your Princely Eminence—I am sending along, lest Your Princely Eminence should not yet have received this little book, a copy of it as current information and with the earnest request that Your Princely Eminence might look graciously upon this highly learned guest on account of his great knowledge and skill, and grant him your gracious protection.[1]

2. ACHILLES PIRMIN GASSER'S LETTER TO GEORG VÖGELIN, PUBLISHED AS A PREFACE TO THE *FIRST ACCOUNT* (2ND EDITION, BASEL, 1541)

To the most learned gentleman Dr. Georg Vögelin of Constance, philosopher and physician, both friend and brother, Achilles Pirmin Gasser of Lindau sends greetings.

Behold, most distinguished Sir, I am sending you, along with the Hercules-stone, this little book. Not only is it something new and unknown to our contemporaries, but also (if I am not mistaken) it will surprise and utterly astonish you even to the verge of disbelief. Georg Joachim Rheticus, master of the liberal arts and sometime professor of mathematics at Wittenberg, my fellow citizen and best friend, a while ago sent it to me from Danzig along with a letter bursting with news about these things. The book certainly departs from the manner of teaching practiced so far. As a whole it may run contrary to the usual theories of the schools and may even sound (as the monks would say) heretical. Nevertheless, what it undoubtedly seems to offer is the restoration—or rather, the rebirth—of a true system of astronomy. For in particular it makes highly evidential claims concerning questions that have long been sweated over and debated all across the world not only by very learned mathematicians but also by the greatest philosophers: the number of the heavenly spheres, the distance of the stars, the rule of the sun, the position and courses of the planets, the exact measurement of the year, the specification of solstitial and equinoctial points, and finally the position and motion of the earth, along with other such difficult matters. And if this man offers to support his reasoning and conclusions with an array of irrefutable demonstration, even if the evidence is only newly adduced, then I do not see who among our learned contemporaries can overthrow, disapprove, or despise his position. For even among those only halfway educated in mathematics, indeed even among (so to speak) their ephemeridical hacks, the business of astronomy (which after all is held to be the most exact science on account of the infallible precision of its circles and computations) today not only is defective in isolated cases of chronological measurement and the observation of celestial motion, but especially falls short of geometry's promise always to produce accuracy.

And so, my dear Georg, thus anticipating our liberation from many astronomical difficulties and the untying of the most tangled knots, I ask you please to read this book carefully right the way through, scrutinize its contents critically, and then recommend what you have scrutinized to all who cherish mathematics, especially those with whom you have close connections. Recommend that they read it too. Not only will this hasten the appearance of a further, fuller account. But also, that rare and almost divine work (whose contents are here adumbrated) will become better known, and a greater stream of requests will reach its author—a man no doubt of incomparable learning and of Herculean, even Atlantean exertion—imploring him to allow the delivery of his whole work to us by means of the persistence, effort, and tireless diligence of my friend, who is deserving of immense respect among contemporary authors.

What I mainly want to promote in this foreword, however, is that you as an unquestionably great expert in the natural sciences should offer your fellow adherents of these most noble disciplines an opportunity whereby the new generation may gratefully prosper and those who are experienced may set truth free—the more liberally, the more fruitfully—in spite of the critical gaze of the vulgar. For you clearly perceive what this announcement requires, and how great is the prize that it promises. And so, with other noble spirits, as is your habit, turn your mind thus to receive and to welcome this little book, lest in future we discover to our sorrow and deep regret that we were utterly robbed and bereft of a most splendid and wholesome banquet—of which indeed this book gives us a rich foretaste—like delicious tender morsels snatched away from hungry mouths.

Farewell, my friend, and for my sake laugh off the opinions of the vulgar regarding this matter, since indeed there is no doubt that this new thing will one day be accepted without bitterness by all educated people as something both agreeable and useful.

Feldkirch, Rhaetia, 1540[2]

3. ACHILLES PIRMIN GASSER'S DEDICATION TO CASPAR TÄNTZL IN HIS *PRACTICA* FOR 1546 (NUREMBERG, 1545)—THE FIRST MENTION OF COPERNICUS IN A WORK PUBLISHED IN GERMAN

To the noble and worthy Caspar Joachim Täntzl of Tratzberg, etc., his most gracious and beloved master, Achilles Pirmin Gasser of Lindau, doctor of natural and medicinal arts, offers his willing service and best regards.

APPENDIX

Noble, worthy, and gracious master, Your Worthiness doubtless still remembers the disputation that you often engaged in with me, not without exceeding amazement, concerning astronomy, while I was in your service in Schwatz last year, and above all the conversation in which—with the help of a small book I had with me, which will eventually be printed—I expressed my desire for a large lodestone, whereby the course of the sun and also the disposition of the firmament (which in the schools we call the Primum Mobile, though we know not where or what it is) would here on earth be rendered calculable and thoroughly perceptible in such a way that no more defects, so frequent until now, should appear.

Moreover, as I then indicated to Your Worthiness, the greatest masters of this art have continuously, for seventeen hundred years, found the movement of the stars and planets rather incongruous and imperfect according to their instruments and calculations, indeed even according to their daily experience. For this reason, one after the other, they always kept on hoping to adjust, improve, and remedy this situation by means of clever contrivances and ingenious speculations, as is evident in Hipparchus, Ptolemy, Al-Battani, Al-Zarqali, Al-Bitruji, Cusanus, Regiomontanus, and finally Werner, with each one correcting the other, now inventing new spheres, then discarding the old ones and thinking up something else, and on and on, with no end of cycles, epicycles, and theoricae—until now so recently, in our own day, also the most learned and wonderful man Dr. Nicolaus Copernicus, off there in Prussia, has taken up the task with such seriousness, diligence, and steadfastness, that for the establishment and restoration of astronomy he has had to lay an utterly and completely new foundation, unheard of before, or rather has been compelled to posit hypotheses not employed by other scholars (namely, that the sun is a light for all creation and stands unmoved in the midst of the whole universe; that this earthly realm together with the other three elements and the circuit of the moon variously courses round between the planets Venus and Mars; and also that the heavens beyond Saturn, in which are seen the fixed stars, all together stand fast and unmoved, with no other spheres encompassing them, etc.), and thus has not only demonstratively proven his doctrine among the mathematicians, and with great pains restored the portrait of Astronomy, but has also immediately been regarded as having perpetrated a heresy, and indeed—by many others incapable of understanding this matter—is already being condemned.

Since now Your Worthiness has for the benefit of this art and sundry other things promised to extract and provide me with a large lodestone from your mine, I have a good will to see progress in this matter and am moved to put into writing these my *Practica* for the coming year 1546, for as Your Worthiness has no meager capacity in astrological predictions to make discriminations and record nature's signs—which, however, must be derived solely from the courses of the planets and their position or placement relative to the other stars—you may easily weigh how very necessary it is that he who can help, advise, and give impetus to such an undertaking should do so, in order that it can actually be brought to fruition.

So I hoped in particular that I might in part accomplish this by means of a large lodestone. I would like therefore to ask Your Worthiness to execute the specified transaction, and to be gracious to accept this my published dedication in your honor, for I remain ever willing, whenever I may, to demonstrate to Your Worthiness my love and service.

May God in heaven be with us, and likewise ever protect your noble and virtuous wife and dear children.

Feldkirch, Monday, July 27, 1545[3]

4. ACHILLES PIRMIN GASSER'S DEDICATION TO RHETICUS IN HIS *PROGNOSTICUM ASTROLOGICUM* FOR 1546 (NUREMBERG, 1545)

To the most learned master of philosophy and the liberal arts, Mr. Georg Joachim Rheticus, most illustrious professor of mathematics at the renowned University of Leipzig—to his friend and fellow citizen—Achilles Pirmin Gasser, medical doctor of Lindau, sends peace and greetings.

I hope you will not instantly contract your brow nor raise your nose in contempt at this effort of mine. But, my Joachim—when so recently you came, unexpectedly and carried by I know not what fate from Saxony back home to us, here to be my dearest guest, it was proper that I should greet you with some choice gift. And just at that time—appropriately, for in my view the timing could not be better—our fruitful Urania, unless of course she miscarried, gave birth to this stricken offspring.

It is indeed a dubious thing and perhaps fit to be abandoned to the wild beasts in the desert. Nevertheless, the love of its mother (though she is not greatly pleased with her progeny) has nurtured and preserved it for so long, that I hope I have not in vain expended my anxious efforts upon it. Thus at last I display it in public, offering and dedicating it to you, if not as a legitimate child for you to nourish, then at least as a deformed lump (as the midwives say), like a frisking kid to which the nanny goat has not yet given birth, in place of the prodigious gift that civility requires I should send you. I implore you, be pleased to accept it, especially since all cultivated men hold that one cannot offer a more fitting gift to a literate man than something literary.

So, by Jove, after praying to the gods that you would arrive here safely, I had nothing else with which, beyond common courtesy, to demonstrate my good will and friendship toward you. Nevertheless, there is another far more powerfully compelling reason why today I dare to burden your outstanding magnanimity with such an ephemeral offering (for one could hardly grant that anybody is honored by such putrid stuff).

For that unremitting passion for the restoration of astronomy continually longs for, desires, implores you, with devotion constant and overflowing, to proclaim with no meager praise that new and paradoxical account of the celestial science, which, since the demise of that most illustrious gentleman Nicolaus Copernicus, your teacher, it is now up to you alone to do. So for us uncultivated little people, having scant aptitude and no Theseus to lead us up the steep terrain, offer an easier introduction to this science, as well as clearer proofs. But above all may you strive to bring forth tables accessible to those of common capacity.

For as no mathematician may dare to refuse or deny not only that the motions of the planets can be determined by a common calculation, to which some riffraff are already taking exception, but also that the equinoctial points (upon which we everywhere rely, though quite insecurely, in our annual prognostications) and the exact and reliable measure of a year, over and above the number of spheres, are not yet observable or fully established, so you have undertaken a work of value, one never to be regretted, and for which all the learned will be forever grateful, if you—turning the accepted hypotheses upside down—can teach those nine muses, first, that the sun, by which we live, stands firm in the center of the universe, whereas the earth, on which we live, moves amongst the planets making its own manifold revolutions, and finally that the highest sphere of the fixed stars encompasses everything and is immobile. Then besides this you undertake both to make good that which is still lacking in the motion of the superior planets, and, in addition, to address all the phenomena or, as they say in the schools, the travails of the planets, with greater ease and clarity.

It remains then for you to set forth with great clarity how this business is to be most carefully and skillfully written up. Or else, if not everything is yet mature enough, then offer us at least for now the rudiments of the new theories.

Thus, by your help we may clearly and simply behold the body of the celestial machine with its earlier kinks hammered out and come to contemplate it unblinkingly in this unaccustomed manner. And so if, as they say, the birds are favorable to your doing that, you shall soar up to the highest stars.

Greetings, and farewell. And as for the printing of these trifles of mine (of which you are chief author), learn from them or rather be cautioned by them that you would be better advised, as regards your own literary work, to aim at publishing it in a more refined form. For I myself occasionally feel like some hasty dog giving birth to blind puppies.*

From Feldkirch, Rhaetia, your native land, July 27, 1545[4]

5. Caspar Brusch's letter to Joachim Camerarius in Leipzig, relating Rheticus's illness (Lindau, last half of August, 1547)

Greetings. Your letter, my most learnèd master Camerarius, reached me on the first of August. I could see from it that you greatly desired to find out about the health of Rheticus and that whole business. So I wrote off to Joachim, who now for three months has been living in Constance and teaching mathematics there. He has halfway regained his former state of health (after having been, while here with me, severely ill), though he is not yet fully recovered. I would have responded immediately to your letter had I not been expecting to receive a letter from Joachim first. I diligently encouraged him to write you a full account regarding this whole business, which I gather he has now done. For he sent me this letter (enclosed with mine), requesting earnestly that I exercise great care in forwarding it to you.

You ask, however, that I write something myself concerning Joachim's health. Of course I would have done so already, except that it would undoubtedly have been more proper for him to explain everything to you himself. For I know that something about an evil spirit has been rumored abroad

---

*Gasser here echoes the ancient proverb "The hasty bitch brings forth blind whelps."

by commercial travelers. Even if this report is not quite false or devoid of substance, still Joachim is taking it very hard, since he is apprehensive that it could somewhat damage his former reputation.

He lay ill here for nearly five months, and every day I would visit, conversing and fellowshipping with him. Throughout this time I made available to him the whole Bible, in both German and Latin, from my library, as well as many of the devotional writings of Luther, Melanchthon, and Cruciger. These he read and reread so diligently that in the end he knew them through and through. However, on particular occasions, with a full heart and most fervent vows, indeed often in tears, he would call upon the Son of God, awaiting deliverance from him alone.

Such are the things I saw and heard daily. And when you see Joachim again, you will behold someone very different from the man you knew in Leipzig, and much changed from the one

> *Who, of heaven, pondered naught but signs, while he,*
> *On earth, in Heavenly Father scarce believed,*
> *Nor any Ruler knowing all we do*
> *Here on this globe, while he beholds from heaven.*

He is completely devoted to his studies and to the sacred readings. Indeed, when he was here, on several occasions he vehemently exhorted us to shun the world's and humankind's egregious complacency, whereby we live day by day disbelieving that there are evil spirits or any avenging furies called forth by our wickedness—lacking all fear of God, and not hesitating to do just whatever we please.

I enjoyed our conversations immensely. Either he would join me in my tutorial room, or else I would visit him, and at such meetings he would speak to me at length concerning such godly themes, or about current events both fortunate and unfortunate, and also offer predictions regarding the outcome of this calamitous war. Truly, he said many things that endeared him to me very much.

His parents, who live in the neighboring town of Bregenz and are exceedingly wealthy as well as quite influential, pressed him fervently to make a pilgrimage to the renowned shrine of St. Eustatius in Alsace (where the papists deliver many from demon-possession). But he would have none of this. It was his conviction instead that he should seek deliverance from Christ alone, the

Son of God, who crushes the serpent's head, and who alone was born to destroy the works of the devil. Accordingly, he earnestly commended himself to the godly churches, requesting their prayers, and also attended Holy Communion with ever deeper self-examination, both here with us and in Ravensburg, to which he and I would accompany each other on horseback.

I write all this to you as to a friend—but also as one might write to a second father—concerning things I would never reveal to anyone with whom I knew they could damage our Joachim in any way. I have written at length, however, both to comply with your request and to show how favorably divine reproof and discipline have transformed one who has always been an object of your benevolence and very dear to you.[5]

6. INVENTORY DRAWN UP BY ANDREAS SIBER, DECEMBER 1, 1551, BY ORDER OF THE LORD MAYOR OF LEIPZIG, OF ITEMS LEFT BY RHETICUS IN OTTO SPIEGEL'S VAULT

1. Furniture
   one small black table
   one small bench
   one ladder
   shelves

2. Scientific equipment
   one armillary sphere in a square glass case

3. Paper of various quantities and formats
   ten bales of paper on a shelf
   one bale and two reams in "median" format
   sixteen "books" in blank "median"
   three bales, nine reams, in "half median"
   fifteen "books" in "half median"
   eighteen "books" in small format paper

4. Maps
   one illuminated map affixed to linen backing
   one "raw" map of "Germania," not colored or affixed
   ten maps of "Deutschland," not affixed

5. Printed works

240 copies of the *Astronomical Tables* (one sheet)

one *New Ephemerides*

1450 copies of the *Canon* (three sheets)

nineteen copies of the "small" *Almanac* (two sheets)

forty-eight copies of the "large" *Almanac* (three sheets)

nine *On Death* (long format, 9 1/4 sheets)[6]

7. FROM PAUL FABRICIUS'S POEM TO RHETICUS FROM VIENNA, 1554

To the most illustrious and learned mathematician Georg Joachim Rheticus: greetings from Paul Fabricius, with the prayer that he might publish his science of triangles.

> *Rheticus, whose labors impel the sky and stars . . .*
> *I send this greeting to you from Danube's banks,*
> *From whose reedy soil stately Vienna rises.*
> *You see this letter, but how I should prefer*
> *You saw its author in person, and he you.*
> *How often I vainly long, with fervent prayers,*
> *To share your presence. How often I desire*
> *Vienna might enfold you and your muses*
> *In her academic embrace. But no, I dream;*
> *And my many prayers are granted no success.*
> *I must greet you from afar. The other way*
> *I may not travel; this way alone stands open.*
> *So let us take the only route we may,*
> *That I be not bereft of your dear favors.*
> *May my letter, first, be welcomed as a friend.*
> *Then may surpassing fruits from your own pen*
> *Go forth, a harvest for many eager hands.*
> *For Triangle has oft desired by your assistance*
> *To be laid open to the learned. She swears—*
> *She implores—by her holy sides that you alone*
> *Should grant this wish. She desires that she be known*
> *And used by all—and judges that you alone*

Can grant this wish. And if you do, your fame
(For you know how great Triangle's power is)
Shall far surpass the fame of other men.[7]

A wall-mounted Nuremberg sundial illustrates how a shadow-casting triangle
can be used to measure part of an astronomical cycle (in this case, a day).

8. FROM RHETICUS'S LETTER TO KING FERDINAND
(KRAKÓW, 1557)

Nicolaus Copernicus, the Hipparchus of our age whom no one has words
enough to praise adequately, was the first to understand the irregularity in
the sphere of the fixed stars . . . In fact, after I had spent about three years
in Prussia, the great old man charged me to carry on and finish what he,
prevented by age and impending death, was himself unable to complete . . .

It must be said that, concerning the positions of the stars in the heavens,
there is much in Ptolemy's account that is not right. To start with, those who

copied his works often ignorantly introduced numerical errors in their transcriptions of tables. And so, because the observations are accurate to scarcely a tenth of a minute even when measured with an armilla, the resulting error can be up to half a degree.

How critical it is to offer a precise account of the starry sphere . . .

So I devoted myself to this immensely important subject at the behest of my master Copernicus, whom I revere as not only my teacher but also my father, honoring and ever striving to please him.

According to Pliny, the first obelisks were established in Egypt.* . . . He also testifies that obelisks are consecrated to the sun-god, which is the meaning of that word in Egyptian. Thus the sun is king and ruler of the heavenly realm, all the other stars being governed by his motions and rhythms. And he is the very eye of the world, by whose light all things are illumined.

Thus by the obelisk alone, all the laws of this heavenly kingdom may be exactly discovered and described. Only the obelisk opens the eyes of artists. By its light we may observe and chart the heavenly motions, seeking by its help fitting proofs and continuously acquiring more useful observations of the motions. Stiborius skillfully admonishes those who come after him when he writes that knowledge of the motions must be supported by new observations. We should not be ever clinging to the old written tables but rather educating ourselves by the heavens themselves.

Therefore the obelisk is no human invention. It was ordained by God the creator not to satisfy human curiosity but to teach God's geometry in heaven and on earth.

The origins of geometry, arithmetic, and astronomy are from the Egyptians, not the Greeks or the Romans. If we are to believe Josephus, since the time of Abraham and the Patriarchs it is from Egypt that mathematics was transplanted into Greece by Plato and into Italy by Pythagoras. Pliny says Pythagoras was also at that time in Egypt, where that obelisk of one hundred twenty-five and a quarter feet was erected which Caesar Augustus later set up in Rome.

The Egyptians call obelisks nature's interpreters, or, better still, nature's

---

*Famous obelisks known as "Cleopatra's Needle" appear today on the Embankment in London and in New York's Central Park. In Rome two huge obelisks, both brought originally from Heliopolis in Egypt, stand at focal points of their respective squares, Piazza del Popolo and Piazza San Pietro.

The world's oldest standing obelisk, located just where Rheticus would have expected: in Egypt, near Heliopolis (the Senusert I Obelisk, sixty-five feet high, dating from ca. 1940 BC).

interpretation. Yet who, beholding this stack of rock, would consider it nature's interpreter, or interpretation? No one, I would think—unless he was like those whose traces were discovered by Aristippus, shipwrecked and beyond hope on the [Rhodian] shore. He bid his unfortunate comrades to be of good cheer, for here he saw signs of humanity: geometrical shapes inscribed in the sand.

The obelisk, then, as I have clearly shown, is philosophy's own device whereby we may determine the positions, periods, and laws of motion of the sun, moon, planets, and other stars. We may use it to investigate stellar motion and all the laws that govern the heavenly realm. And with its help we may construct, establish, and promote astronomy, geography, and that part of physics concerned with the effects of the stars and so pertinent to the life of humankind.[8]

9. LETTER FROM RHETICUS TO PAUL EBER IN WITTENBERG (KRAKÓW, OCTOBER 1561)

To the most illustrious and learned Mr. Paul Eber, Doctor of Theology, Pastor of Wittenberg, my old friend:

Greetings. The person delivering this letter, Jodok Rab, would like to study with you in Wittenberg. He has already studied with modest success in Strasbourg and elsewhere in Germany. As I think him a decent young fellow, I have no hesitation in recommending him to you—especially if it might lead to more frequent correspondence between you and me.

I know that your meditations on the invisible Heaven do not keep you from considering the visible heavens, and that you are observing the workings of the stars. As some (for example, [Sebastian] Münster) seek to assemble genealogies of princes, so I would like to see you have one of your people undertake the business of recording their generations in an orderly fashion. I wanted to do so myself using your *Calendar,* but other obligations are keeping me from it, and there is no one here whom I can get to do it, even if I could afford it. When it comes to mathematics, this place is a wasteland.

As for Carion's *Chronicle:* Who will revise and update it if you do not? I know what I am talking about. Take care not to deprive us, and posterity, of your accomplishments.

I commend this young man to Mr. Peucer and to others who may be able to assist him. Farewell.

Kraków, October 8, 1561

Please greet our mutual friends affectionately in my behalf.

Yours,
Joachim Rheticus[9]

10. Letter from Rheticus to Paul Eber in Wittenberg (Kraków, November 1562)

To the most illustrious and learned Mr. Paul Eber, Doctor of Theology, Pastor of Wittenberg, my dearest friend:

Greetings, and thanks for not letting the courier leave without your letter to me. I desire nothing so much as one single friend whom I could talk to occasionally. Medicine readily gives me enough money and prestige. I avoid the upper crust because I do not want to waste my time. It is better to live out one's life among the Papists than among the Calvinists, or whoever they are. Therefore I am greatly looking forward to seeing your little book [on the Sacrament]. Our people are bringing out some of Melanchthon's letters, but I have not seen any of this material. These days there is so much writing and so little doing! May God have mercy.

Here, the Pope seems up to something, and I fear the fight between Constantine and Licinius—I mean that the whole world will have to turn either Papist or Calvinist. If only we could just be Christians! However, the disputes these days are no longer about the sacrament but about the divinity and humanity of Christ. Why cannot all intelligent people merely strive to live aright, and to do what is pleasing to God and useful to their fellow men?

November 1, 1562

Yours,
Joachim Rheticus[10]

11. Rheticus's "manifesto"—an excerpt from his letter to Petrus Ramus in Paris, 1568

The following works, Ramus, are partly completed, partly in process. First, because in my judgment astronomy and geography must be examined right from their foundations, I have written three sections on constructing a canon of the science of triangles, and have worked out the roots of the canon by means of algebra. In addition I have put together a small canon for everyday use, with distances from the center, or opposite sides, reckoned at 100,000. The larger canon for more precise geometrical research indeed requires 10,000,000,000.

But I would like the first series of the canon to be still further broken down into the following Arabic numbers: 100,000; 10,000,000,000; 1,000,000,000,000,000—and all of these not only for distances about the circumference and in degrees, but also calculated to tenths of minutes. This has been a dozen years' work, and that with several computational workers continuously on my payroll.

Secondly, I have composed a new collection of chapters on the science of spherical triangles both with right angles and without right angles. In addition I am proposing a volume of ten lessons on plane triangles, to which I likewise hope to add a further nine chapters, God granting life.

Thirdly, these are to be followed by a work of nine chapters concerning celestial phenomena. In it I shall first demonstrate the method of taking observations, by means of which we can determine precisely the positions of the stars, lights, planets, comets, and everything else that can be observed on high. I shall then follow the astronomical work accordingly with a geographical one. In this work I have resolved—almost as if it were a sport—to present the various applications of numerous new astronomical and geographical tables.

However, I show how, merely by means of the science of triangles which I teach, and by means of my canon of triangles without any other tables, everything can be quite easily calculated. If commercial accountants, as it seems, expend such effort and care in their business calculations—"practica," as the Italians call them—along with all kinds of other computations, then why should not we, in pursuing higher things, likewise also devise powerful and precise means for making our calculations? In this work, I introduce examples from all the lesser geometrical works of Proclus.

Fourthly, I shall now undertake that work which you too have in mind: freeing the science of astronomy from its hypotheses and pursuing it solely based on observations. Yet if only we possessed observations from every epoch and understood how the methods compared with those handed down to us (these being in my opinion the same as those used by the first practitioners of this science)—and if we could perform computations by means of such tables as did not require continuous emendation!

To this work I shall append tables of unequal motions, from which I can determine planetary position and all the other phenomena as easily as if I were reading them from ephemerides.

In all these matters I am much more averse to Ptolemy than you are to Euclid; I chastise him more severely than you do Euclid. For just as the huts that children build out of mud and sand compare with the buildings of Vitruvius or the palaces of Rome in its glory, so the majestic edifice of Ptolemy—one were better to call it a magnificent ruin—compares with the true and reliable science of stellar motion, which you could also call an Egyptian astronomy. For the Egyptians with their radii (which the Greeks out of ignorance call obelisks) pursued this science blessed with quite divine understanding.

Finally, illustrious Ramus, I am composing a German astronomy for my Germans. As far as the effects of the stars are concerned, I am assembling astrological compendia. Here too I have almost in hand a new method for conducting natural science, derived solely from the investigation of nature itself, quite independently of all the ancient writings.

Likewise in the science of medicine. And since I take great delight in chemistry, I have dug down to the foundations of this science, with the result that I have sketched out seven chapters concerning it.

Such and so weighty are the things I am working on—and for which, so far, the practice of medicine, my Maecenas, has born my expenses.[11]

NOTES

When a quoted English translation was borrowed from another scholar, the endnote names him or her. Otherwise, translations are my own.

ABBREVIATIONS

ACTA — *Acta Rectorum Universitatis Studii Lipsiensis* (1524–58), edited by Friedrich Zarncke (Leipzig: Tauchnitz, 1859)

BMSTR — Karl Heinz Burmeister, *Georg Joachim Rhetikus, 1514–1574: Eine Bio-Bibliographie*, 3 vols. (Wiesbaden: Guido Pressler, 1967–68)

BOTC — *The Book of the Cosmos: Imagining the Universe from Heraclitus to Hawking*, edited by Dennis Danielson (Cambridge, MA: Perseus, 2000)

Cop*MW* — Nicholas Copernicus, *Minor Works*, edited by Pawel Czartoryski, translated by Edward Rosen (Kraków: Polish Scientific Publishers, 1985)

DUDITH — Andreas Dudithius, *Epistulae*, 5 vols., edited by Tiburtio Szepessy et al. (Budapest: Akadémiai Kiadó, 1992–2002)

GINGR — Owen Gingerich, *An Annotated Census of Copernicus' "De Revolutionibus"* (Leiden: Brill, 2002)

Kepler*GW* — Johannes Kepler, *Gesammelte Werke*, edited by Max Caspar (Munich, C. H. Beck, 1938–)

LUTHER — *D. Martin Luthers Werke: kritische Gesamtausgabe*, part 1 (Weimar: H. Böhlau, 1883–)

MCTHN   *Corpus Reformatorum, Philippi Melanchthonis Opera,* edited by
C. G. Bretschneider and H. E. Bindsell, 28 vols. (Halle:
Schwetschke, 1834–60)

Montfort   *Montfort: Vierteljahresschrift für Geschichte und Gegenwart
Vorarlbergs*

NCGAR   *Nicolaus Copernicus Gesamtausgabe,* edited by H. M. Nobis et
al., Vol. VIII/1 (*Receptio Copernicana*) (Berlin: Akademie
Verlag, 2002)

PROWE   Leopold Prowe, *Nicolaus Coppernicus,* 2 vols. (1883–84;
reprint, Osnabrück: Zeller, 1967)

RosenCSR   Edward Rosen, *Copernicus and the Scientific Revolution* (Mal-
abar, FL: Krieger, 1984)

RosenTCT   *Three Copernican Treatises,* translated by Edward Rosen, 3rd
ed. (New York: Octagon Books, 1971)

PROLOGUE: NO RHETICUS, NO COPERNICUS

1. Philipp Melanchthon to Mithobius, Oct. 16, 1541; MCTHN, vol. 4, col. 679.

2. Rheticus's letter to King Ferdinand (Kraków, 1557); BMSTR, 3:139.

3. The request was contained in a letter Copernicus received from Cardinal Nicolaus Schönberg in 1536. Although he did not respond to it at the time, Copernicus took this letter so seriously that he kept it and had it printed as a headnote to *The Revolutions* in 1543. See the illustration on p. 111.

4. According to the opening page of his own *First Account,* Rheticus's health was such that Tiedemann Giese invited him, together with Copernicus, to come for a period of "rest from [his] studies"; PROWE, 2:295.

5. The scenario narrated in this prologue is based principally on two letters by Copernicus's best friend, Tiedemann Giese. In the first letter, written to Georg Donner on December 8, 1542, Giese—acknowledging news from Donner concerning Copernicus's failing health—urged Donner to look after their friend during his illness. In the second letter, sent to Rheticus on July 26, 1543, Giese recounted details of Copernicus's illness and death. My assumptions are (a) that Donner did respond to Giese's request to care for their friend, and (b) that Giese's knowledge of the circumstances surrounding the death of Copernicus was based on information he had continued to receive from Donner. The fact that Rheticus provided Donner with a personally

signed copy of Copernicus's treatise is consistent with these assumptions, as is the fact that Donner sent a copy of *The Revolutions* to Duke Albrecht of Prussia and then engaged in further correspondence with the duke concerning the gift (July 28 and August 3, 1543). All of the correspondence here referred to is collected in Franz Hipler, *Spicilegium Copernicanum* (Braunsberg, 1873), pp. 349–356. Other inferred details regarding Copernicus on his deathbed are based on my own experience attending an almost seventy-year-old man in his last days.

6. Paraphrase of Adolf Müller, "Der Astronom und Mathematiker Georg Joachim Rheticus," *Vierteljahrsschrift für Geschichte und Landeskunde Vorarlbergs*, new series, 1 (1917): 1–46 (p. 41): "Ohne Rheticus kännten wir keinen Kopernicus." Compare GINGR, p. xii: "[Copernicus] would never have seen his work printed except for the intervention of a young professor of astronomy from Lutheran Wittenberg, Georg Joachim Rheticus"; and Edward Rosen, in *Isis*, 61.1 (Spring 1970): 137: "Is it going too far to claim that without Rheticus, no Copernicus; without Copernicus, no moving earth; and without geodynamic astronomy, no modern science? If so, the relations between Rheticus and Copernicus are of utmost importance to the history of science."

1. PATRONS AND A POET

1. Karl Heinz Burmeister, *Kulturgeschichte der Stadt Feldkirch* (Sigmaringen: Jan Thorbecke, 1985), p. 163.

2. Cassiodorus (ca. 490–ca. 583), *Institutiones*.

3. Rheticus, dedicatory epistle to Heinrich Widnauer, in *Orationes duae* (Nuremberg, 1542); reprinted in BMSTR, 3:49–51, also in PROWE, 2:382–386.

4. This may be more than a mere literary allusion. Iserin's case or some confused account of it might in fact have given rise to a version of the Faust legend known as *Doktor Fustes in Feldkirch*. See Erich Somweber, "Der Zauberer und Hexenmeister Dr. Georg Iserin von Mazo," *Montfort*, 1968, p. 86.

5. *Die Peinliche Gerichtsordnung Kaiser Karl V. von 1532 (Carolina)*, edited by Gustav Radbruch (Stuttgart: Reclam, 1962), article 109. One of the weaknesses of this view is the fact that this code did not officially come into effect until 1532, four years after Iserin's execution.

6. "Acta unndt Gericht über Georg Iserin," Stadtbibliothek Feldkirch manuscript sig. Akt 66 (46 pages). The main case against seeing Iserin's crime principally as sorcery is put by Manfred Tschaikner, "Der verzauberte Dr. Iserin," *Kulturinformationen Vorarlberger Oberland* 2 (1989): 147–151.

7. See the account by Karl Heinz Burmeister, "Neue Forschung über Georg Joachim Rhetikus," *Jahrbuch des Vorarlberger Landesmuseumsvereins,* 1974-75 (Bregenz: Freunde der Landeskunde, 1977), pp. 37–47.

8. Somweber, "Der Zauberer," p. 83.

9. BMSTR, 1:13.

10. On Gasser, see Karl Heinz Burmeister, *Achilles Pirmin Gasser, 1505–1577: Arzt und Naturforscher, Historiker und Humanist,* 3 vols. (Wiesbaden: Guido Pressler, 1970–75).

11. Valentin Otto, from the preface ("Ad candidum Lectorem") to his and Rheticus's *Opus palatinum de triangulis* (Neustadt, 1596).

12. Stefan Oehmig, "Wittenberg als Universitäts- und Studentenstadt," in *Wittenberg als Bildungszentrum, 1502–2002,* edited by Peter Freybe (Lutherstadt Wittenberg: Drei Kastanien Verlag, 2002), pp. 36, 34–35.

13. Luther is often cited as an early opponent of Copernicanism on the strength of a few comments he is reported to have made over dinner dismissing the idea of a moving earth. Such dismissal is not too surprising when one considers that sixty years after Rheticus succeeded in having *The Revolutions* published, there were still, by one scholarly estimate, no more than ten certifiable Copernican astronomers in all of Europe. See Robert S. Westman, "The Astronomer's Role in the Sixteenth Century: A Preliminary Study," *History of Science* 18 (1980): 106. In recent correspondence Westman informed me he is inclined to stand by this estimate of numbers. His published list is as follows: "Thomas Digges and Thomas Hariot in England; Giordano Bruno and Galileo Galilei in Italy; Diego de Zuñiga in Spain; Simon Stevin in the Low Countries; and, in Germany, the largest group—Georg Joachim Rheticus, Michael Maestlin, Christoph Rothmann, and Johannes Kepler." Peter Barker recently presented strong evidence suggesting that Rothmann in fact recanted from his Copernicanism: "How Rothmann Changed His Mind," *Centaurus* 46 (2004): 41–57. One further possible candidate for the list is the Englishman William Gilbert; see Howard Margolis, *It Started with Copernicus: How Turning the World Inside Out Led to the Scientific Revolution* (New York: McGraw-Hill, 2002), chap. 5. To these, in fairness, should

probably be added the names of Achilles Gasser and Georg Vögelin (on account of their provision of affirmative front matter for the 1541 edition of the *First Account*), and of Valentin Otto, for whom the evidence is strong if circumstantial. If one then includes the names of Copernicus himself and of his friend and encourager Tiedemann Giese, the list may stand at fifteen—still not a large number.

For an excellent review of responses to Luther's critical comment (made four years before Copernicus's work was published), see Donald H. Kobe, "Copernicus and Martin Luther: An Encounter between Science and Religion," *American Journal of Physics* 66.3 (March 1998): 190–196.

14. From *The Assayer,* in *Discoveries and Opinions of Galileo,* translated by Stillman Drake (New York: Doubleday, 1957), pp. 237–238. For an excellent and thorough introduction to the "two books" theme, see Kenneth J. Howell, *God's Two Books: Copernican Cosmology and Biblical Interpretation in Early Modern Science* (Notre Dame, IN: University of Notre Dame Press, 2002), especially chaps. 1–3.

15. Martin Luther, preface to *Symphoniae jucundae* (1538); LUTHER, 50:372–373.

16. Galileo, *Sidereus Nuncius,* selection translated in *BOTC,* p. 150.

17. Johannes Kepler, *The Harmony of the World,* translated by E. J. Aiton, A. M. Duncan, and J. V. Field (Philadelphia: American Philosophical Society, 1997), p. 254.

18. Many details in this and the following two paragraphs are drawn from Stefan Rhein et al., *Philipp Melanchthon,* 2nd ed. (Wittenberg: Drei Kastanien Verlag, 1998), pp. 12–22.

19. Melanchthon to Erasmus Ebner, July 7, 1542; MCTHN, no. 2514. (Letters are cited by no. assigned to them in MCTHN.)

20. One subsequent record—Andreas Sennert's *Athenae: Inscriptiones Wittebergenses* (Wittenberg, 1655), p. 93—indicates 1541 as the year of Rheticus's installation. However, he performed the duties of this professorship at least as a lecturer from 1536 and was referred to as a professor well before 1541 by no less an authority than Melanchthon. (MCTHN, no. 1740, dated Oct. 15, 1538, tells Camerarius that Rheticus "est Professor Mathematum apud nos.") The presumption that Rheticus held his professorship in the late 1530s is newly supported by the discovery of a manuscript of Rheticus's lecture on arithmetic from August of 1536. See *Die Arithmetik-Vorlesung*

des Georg Joachim Rheticus, Wittenberg 1536, edited by Stefan Deschauer; *Algorismus: Studien zur Geschichte der Mathematik und der Naturwissenschaften,* issue no. 42 (Augsburg: Erwin Rauner Verlag, 2003).

21. Rheticus's lecture appears in *Melanchthon: Orations on Philosophy and Education,* edited by Sachiko Kusukawa, translated by Christine F. Salazar (Cambridge: Cambridge University Press, 1999), pp. 90–97. Quotations are from this translation.

22. For example, book 2, epigram 76, consists of the following erotic dialogue:

> You tell me oft, "If you but saw me naked,
> > Then not my calves alone would surely please you;
> My tender unsucked nipples you would praise;
> > You'd praise as well my breasts, more white than snow."
> "Yet something else invites my approbation."
> "My thighs?" you ask. "No, not your thighs," I answer.
> "My thighs displease you, then?" "No, they are lovely.
> > But one thing lovelier still lies hid nearby."

Lothar Mundt, *Lemnius und Luther: Studien und Texte zur Geschichte und Nachwirkung ihres Konflikts (1538/39),* 2 vols. (Berne: Peter Lang, 1983), 2:86.

23. Ibid., 1:27.
24. LUTHER, 46:433–439.
25. Mundt, 1:32.
26. Ibid., 2:337. In translation, the poem reads:

> How well your theme and poem suit you, Lemmy.
> For shit's your theme, your poetry's the shits,
> And shitty poems befit the shithead Lemmy.
> For what but shit so suits a shitty poet?
> Pity the prince you praise in shitty verses,
> His reputation spattered by your shit.
> You strain to move your bowels, to shit a mountain,
> But, O Shit-Poet, your turds amount to nothing.
> And should your trespass find its fit reward,
> Your sorry, shitty corpse will feed the ravens.

27. Ibid., 2:151. In translation, the poem reads:

> You cry out, pained by your own dysentery!
> The excrement is yours, not someone else's;
> While calling others shitty, it is you
> Who prove the shitter, and how your shit abounds.
> Where formerly your crooked mouth spewed madness
> It's now your arse from which you vent your spleen.
> Your madness has no need to pass your throat,
> When it can be expelled from your behind.
> Nor need your vileness spill forth from your lips:
> It can be farted forth another way.
> I think, though, that you'd rather burst your guts
> Than vent such violence out your nether sphincter.

28. Ibid., 2:182.

2. ROUNDABOUT ROAD TO FRAUENBURG

1. The biography is that of Pierre Gassendi (1592–1655), abridged in *Regiomontanus on Triangles*, translated by Barnabas Hughes (Madison: University of Wisconsin Press, 1967), p. 11.

2. Maximilian Curtze, "Der Briefwechsel Regiomontan's mit Giovanni Bianchini, Jacob von Speyer und Christian Roder," *Abhandlungen zur Geschichte der mathematischen Wissenschaften*, 12 (1902): 327; translated in Rosen*CSR*, p. 171.

3. Schöner's unique copy of this map, long thought lost, was rediscovered in 1901, and in 2001 the U.S. Library of Congress agreed to purchase it at a cost of ten million dollars. See "America's Birth Certificate," www.loc.gov/loc/lcib/01078/discovering.html.

4. "Nemini tamen triangulos nostros praetereunti astror[u]m disciplina satis agnoscetur." The translation is that of Hughes, in *Regiomontanus on Triangles*, p. 27.

5. See "Johannes Petreius, Nuremberg Publisher of Scientific Works, 1524–1550, with a Short-Title List of His Imprints," in *Homage to a Bookman: Essays on Manuscripts, Books and Printing*, edited by Helmut Lehmann-Haupt (Berlin: Mann, 1967), pp. 146–162.

6. Translated by N. M. Swerdlow, "Annals of Scientific Publishing: Johannes Petreius's Letter to Rheticus," *Isis* 83.2 (June 1992): 273–274.

7. Ibid., p. 273.

8. Rheticus to King Ferdinand, 1557; BMSTR, 3:136.

9. See Heinz Balmer, *Beiträge zur Geschichte der Erkenntnis des Erdmagnetismus* (Aarau: Sauerländer Verlag, 1956), pp. 287–292.

10. BMSTR, 3:46.

11. See http://www-gap.dcs.st-and.ac.uk/~history/Mathematicians/Apianus. html. See also Christoph Schöner, "Apian, Peter," in *Die Deutsche Literatur,* series 2, section A, "Autorlexikon" (Stuttgart-Bad Cannstatt: Frommann-Holzboog, 2001), no. 257.

12. BMSTR, 3:191.

13. From the first page of Rheticus's *First Account (Narratio prima,* Danzig, 1540); translated in Rosen*TCT,* p. 109. The letter from Posen, of which Rheticus makes mention, has not survived.

3. *VITA COPERNICI*

1. Rosen*CSR,* p. 168.

2. Dantiscus to Giese; in Ludwik Antoni Birkenmajer, *Mikołaj Kopernik* (Kraków, 1900), p. 395, no. 13. Dantiscus's Latin refers to Anna Schilling as "scortum suum."

3. The "heretic" referred to by Dantiscus was Alexander Scultetus. Dantiscus to Giese, July 5, 1539; in Rosen*CSR,* pp. 158–159.

4. "Against Werner" quoted from Rosen*TCT,* pp. 93–106; cf. Latin original, in PROWE, 2:172–183.

5. Rosen*TCT,* p. 374.

6. PROWE, 1(2):274. Widmanstetter is also known as Widmanstadt.

7. Ibid., 1(1):375.

8. Ibid., 1(1):80.

9. Karl Heinz Burmeister, private conversation, March 20, 2003.

10. "His carneus diabolus"; PROWE, 1(1):376–377.

11. The full letter is printed in ibid., 2:51.

12. Cop*MW,* p. 38.

13. Nobody calls it an outstanding translation. Perhaps typical is J. L. Heilbron's dry comment in *The Sun in the Church: Cathedrals as Solar Observatories* (Cambridge, MA: Harvard University Press, 1999), p. 7: "The

choicer Greek authors had long since found modern [i.e., Renaissance] editors, luckily for them."

14. *Collationes in Hexameron*, 12.14; in *The Works of Bonaventure*, translated by José de Vinck (Paterson, NJ: St. Anthony Guild Press, 1970).

15. Rosen*TCT*, pp. 352–353.

16. Cop*MW*, p. 187.

17. Ibid., pp. 176–177; cf. PROWE, 2:33.

18. From *The Revolutions*, bk. 1, chap. 10, as translated in *BOTC*, p. 117.

19. Cop*MW*, p. 192.

20. *Narratio prima;* Rosen*TCT*, pp. 146, 147.

21. Cop*MW*, pp. 190–191.

22. *The Revolutions*, bk. 1, chap. 9, as translated in *BOTC*, p. 115.

23. *The Presocratics*, edited and translated by Philip Wheelright (New York: Odyssey Press, 1966), p. 184.

24. Translated and quoted by Rosen*TCT*, p. 368.

25. Cop*MW*, pp. 343–348.

4. WHAT RHETICUS KNEW

1. *Commentariolus;* in Rosen*TCT*, p. 57.

2. Translations from the *Commentariolus* are adapted from Rosen*TCT*, in consultation with the one by Noel M. Swerdlow, "The Derivation and First Draft of Copernicus' Planetary Theory: A Translation of the *Commentariolus* with Commentary," *Proceedings of the American Philosophical Society* 117 (1973): 423–512, as well as with the original Latin text, in PROWE, 2:184–202.

3. Ptolemy, *Almagest;* excerpt reprinted in *BOTC*, p. 72.

4. This value is actually taken from Ptolemy. See Albert Van Helden, *Measuring the Universe: Cosmic Dimensions from Aristarchus to Halley* (Chicago: University of Chicago Press, 1985), p. 27. It is not known exactly what Copernicus considered these distances to be.

5. Alfred Romer, "The Welcoming of Copernicus's *De revolutionibus:* The *Commentariolus* and Its Reception," *Physics in Perspective* 1 (1999): 181.

5. COPERNICAN SUNRISE

1. "Against Werner" quoted from Rosen*TCT*, pp. 93–106; cf. Latin original, in PROWE, 2:172–183.

2. All quotations in this paragraph are from Rheticus, *First Account*, in Rosen*TCT*, pp. 192–194.

3. Rosen*TCT*, pp. 188–196. All quotations of the *Encomium* are taken from Rosen*TCT*, in consultation with the Latin, which appears in PROWE, 2:367–377.

4. See *Kepler's Conversation with Galileo's Sidereal Messenger*, translated by Edward Rosen (New York: Johnson Reprint Corporation, 1965), pp. 43–46.

5. Rosen*TCT*, p. 177.

6. Ibid., p. 137.

7. Ibid., p. 138.

8. See, among others, Ian G. Barbour, *Myths, Models and Paradigms* (New York: HarperCollins, 1976); *Metaphor and Analogy in the Sciences*, edited by Fernand Hallyn (Dordrecht: Kluwer, 2000), especially the chapter by Gérard Simon, "Analogies and Metaphors in Kepler," pp. 71–82.

9. *BOTC*, p. 117. The term *symmetry* in this passage carries the literal sense of "common measure" that was so important to Copernican astronomical thought.

10. For an exposition of this view and its implications for the interpretation of Copernicus, see Dennis Danielson, "The Great Copernican Cliché," *American Journal of Physics* 69.10 (Oct. 2001): 1029–1035.

11. Giovanni Pico, *Oration on the Dignity of Man*, in *The Renaissance Philosophy of Man*, edited by Ernst Cassirer et al. (University of Chicago Press, Chicago, 1948), p. 224. The original phrase is "excrementarias ac foeculentas inferioris mundi partes"; *Opera omnia Ioannis Pici* (Basel, 1493), p. 314.

12. Rosen*TCT*, p. 139.

13. Adapted from ibid., p. 140. The original of the final phrase reads: "pleraque Veterum Solis ἐγχώμια tanquam poëtica negligebamus" (PROWE, 2:321). The importance of this sentence was not lost on Rheticus's earliest readers. Whoever prepared the second edition of the *First Account* in 1541 (perhaps its publisher, Robert Winter) provided an index dominated by proper names but including an entry for poetry (*poëtica*), which refers to this poetic setting forth of the sun's rule.

14. Rosen*TCT*, p. 151.

15. Adapted from ibid., p. 143; original text in PROWE, 2:324.

16. Rosen*TCT*, pp. 151, 144.

17. Galileo, *Sidereus Nuncius;* excerpted in *BOTC*, p. 150.

6. GOD'S GEOMETRY IN HEAVEN AND ON EARTH

1. BMSTR, 2:45.

2. Quoted in ibid., 1:47; also Hipler, *Spicilegium Copernicanum*, pp. 351–352. For a translation of the whole letter, see the appendix, item 1.

3. See chapter 2, note 6.

4. Letter from Gemma Frisius to Dantiscus, July 20, 1541; in Henry de Vocht, *John Dantiscus and His Netherlandish Friends as Revealed in Their Correspondence, 1522–1546* (Louvain: Librairie Universitaire, 1961), pp. 345–346.

5. See Karl Heinz Burmeister, "Der Konstanzer Arzt Dr. med. George Vögelin (1508–1542), ein früher Anhänger des Kopernicus," *Archiwum Historii i Filozofii Medycyny* (Kraków) 62.1–2 (1999): 97–104. See also Burmeister, "Georg Joachim Rhetikus—ein Bregenzer?" *Montfort* 57.4 (2005): 308–327.

6. BMSTR, 3:15–18. For a fuller discussion of Gasser's response, see Dennis Danielson, "Achilles Gasser and the Birth of Copernicanism," *Journal for the History of Astronomy* 35 (2004): 457–474. For a translation of the whole letter, see the appendix, item 2.

7. A German translation of Vögelin's Latin poem appears in George Joachim Rhetikus, *Erster Bericht*, translated by Karl Zeller (Munich and Berlin: Oldenbourg, 1943), p. 26.

8. Samuel Y. Edgerton, Jr., *The Heritage of Giotto's Geometry: Art and Science on the Eve of the Scientific Revolution* (Ithaca: Cornell University Press, 1991), pp. 223–253.

9. BMSTR, 3:139, 134.

10. All quotations of this work are taken from F. Hipler, "Die Chorographie des Joachim Rheticus, aus dem Autographon des Verfassers," *Zeitschrift für Mathematik und Physik* 21 (Leipzig, 1876): 125–150. The text is reprinted under the title *Chorographia Tewsch*, in *NCGAR*, pp. 77–88. For more on chorography, see Gerald Strauss, *Sixteenth-Century Germany: Its Topography and Topographers* (Madison: University of Wisconsin Press, 1959).

11. Hipler, "Chorographie," p. 145.

12. Ibid., p. 149.

13. See William Gilbert, *De Magnete* (1600), translated by P. Fleury Mottelay (1893; reprint, New York: Dover, 1958); and Howard Margolis's

discussion of Peregrinus, Gilbert, and the spherical lodestone (*terrella*, or "little earth"), in *It Started with Copernicus*, pp. 145–152.

14. BMSTR, 3:40.

15. Ibid., 3:138; for the full letter, see the appendix, item 8.

7. COPERNICAN AFTERLIFE

1. Eber to Melanchthon; MCTHN, no. 2194.

2. Melanchthon to Mithobius, Oct. 16, 1541; MCTHN, no. 2391.

3. BMSTR, 3:43; "omnium humanorum operum longe pulcherrimum."

4. Ibid., 3:43–44.

5. Quotations from the orations are taken from *Melanchthon: Orations on Philosophy and Education*, edited by Kusukawa, chaps. 13 and 16 (although this publication fails to make it clear that these orations are by Rheticus, not Melanchthon). The original orationes (*Orationes duae*) may be consulted in NCGAR, pp. 106–117.

6. Quotations from the dedication of *On the Sides and Angles of Triangles* are free translations of the Latin text reproduced in BMSTR, 3:45–47 and PROWE, 2:378–381.

7. Rheticus to Widnauer, 1547; in BMSTR, 3:50.

8. MCTHN, no. 2484.

9. Ibid., no. 2514.

10. Ibid., no. 2526.

11. Ibid., no. 2607.

12. See Burmeister, "Neue Forschung," p. 41; and Burmeister, "Rheticus— ein Bregenzer?" p. 311.

8. WELCOME TO LEIPZIG

1. MCTHN, no. 2577.

2. *Die Matrikel der Universität Leipzig*, edited by Georg Erler (Leipzig: Giesecke and Devrient, 1897), 2:669.

3. See Melancthon's comments to Camerarius concerning "our Joachim," MCTHN, no. 2489.

4. The account here and in the following paragraph is based on Erler, *Die Matrikel*, p. 671; see also p. 196.

5. Fragments of two 1540 letters from Osiander to Rheticus, the first dated March 13, 1540, the second undated; in Rosen*CSR*, pp. 192–193. See Rosen's overall account of the Rheticus-Osiander relations, with documentation, pp. 192–212.

6. Ibid., p. 193.

7. Ibid., p. 194.

8. Rheticus, *De motu terrae*, pp. 1, 12, 40. This tract was published as an appendix to David Gorlaeus, *Idea Physicae* (Utrecht, 1651). The original text and literal English translation are printed in Reijer Hooykaas, *G. J. Rheticus' Treatise on Holy Scripture and the Motion of the Earth* (Amsterdam: North-Holland, 1984). The complete Latin text is also reprinted in *NCGAR*, pp. 57–73. For deeper discussion of the tract than is possible here, see Howell, *God's Two Books*, pp. 255–267. See also Noel Swerdlow's account of how his skepticism about Hooykaas's attribution of *De motu terrae* to Rheticus was overcome: *Journal for the History of Astronomy* 17 (1986): 134.

9. See Hooykaas, *Rheticus' Treatise*, pp. 17–19, for the account of Hooykaas's discovery.

10. Giese to Rheticus, July 26, 1543; in Rosen*CSR*, p. 167.

11. Copernicus, *On the Revolutions*, translated by Edward Rosen (Kraków: Polish Scientific Publishers, 1978), p. xvi.

12. *De motu terrae*, p. 14.

13. These translations are adapted from Rosen*CSR*, p. 167; cf. the Latin text in BMSTR, 3:54–55.

14. Birkenmajer, *Kopernik*, p. 403; translation by Rosen*CSR*, p. 197. See this and the following pages in Rosen*CSR* for a fuller account of the roles of Petreius and Osiander with respect to the "Ad lectorem" Rosen's account is then amended in GINGR, pp. 219–221.

15. Translation based on Rosen*CSR*, p. 198, but amended in light of both the Latin text and Gingerich's translation, GINGR, pp. 220–221. The Maestlin text also appears in Ernst Zinner, *Entstehung und Ausbreitung der Copernicanischen Lehre*, 2nd ed. (Munich: Beck, 1988), p. 453.

16. See GINGR, pp. 135–136.

17. See Westman's discussion, "The Melanchthon Circle, Rheticus, and the Wittenberg Interpretation of the Copernican Theory," *Isis* 66 (1975): 183–186.

18. Peter Barker, "Constructing Copernicus," *Perspectives on Science* 10.2 (2002): p. 220.

9. GAMBLING ON CARDANO

1. See his *Liber de ludo aleae (Book on Games of Chance)*, 1526; in Girolamo Cardano, *Opera omnia* (Lyons, 1663; reprint, New York: Johnson Reprint Corporation, 1967), 1:262–276.

2. *Libelli duo* (Nuremberg, 1543); *Opera*, 5:490: "Ioachimus Georgius noster."

3. Rheticus himself hints at this connection in his retrospective 1550 letter to Komerstadt, referring to Italy as a place "where for some time as a boy I lived with my family" (*ubi puer aliquando cum meis fueram*); BMSTR, 3:108.

4. Preface "To the Reader," *Ephemerides* (Leipzig, 1550).

5. Letter from Gemma Frisius to Dantiscus, July 20, 1541; in de Vocht, *John Dantiscus*, pp. 345–346.

6. The titles of the German and Latin versions are, respectively: *Practica auff das M.D.XLVI. Jar*, and *Prognosticum Astrologicum ad annum Domini M.D.XLVI*. See Karl Heinz Burmeister, "'Mit subtilen fündlein und sinnreichen speculierungen...': Die 'Practica auff das M.D.XLvj. jar' des Achilles Pirmin Gasser im Umfeld zeitgenössischer Astrologen," *Montfort* 55.2 (2003): 107–120.

7. Further discussion of both prefaces written by Gasser on July 27, 1545, is provided in Danielson, "Achilles Gasser and the Birth of Copernicanism," pp. 457–474. The Latin preface is reprinted in BMSTR, 3:68–69, and the German in Burmeister's article in *Montfort*, cited in the previous note. For full English translations, see the appendix, items 3 and 4.

8. From *Libelli quinque* (Nuremberg, 1547); in *Opera*, 5:491; translation from Anthony Grafton, *Cardano's Cosmos: The Worlds and Works of a Renaissance Astrologer* (Cambridge, MA: Harvard University Press, 1999), p. 92.

9. Cardano, *Artis magnae* (Nuremberg, 1545), verso of the title page. Quoted here is Bruce Wrightsman's translation of the letter, from "Andreas Osiander's Contribution to the Copernican Achievement," in *The Copernican Achievement*, edited by Robert S. Westman (Berkeley: University of California Press, 1975), pp. 231–232.

10. Cardano, *Opera*, 5:85.

11. Grafton, *Cardano's Cosmos*, pp. 94–96.

12. Burmeister, "'Mit subtilen fündlein . . .'," pp. 116–119. See Cardano's *De vita propria*, chap. 15, "De Amicis, atque Patronis"; *Opera*, 1:12–13.

13. Grafton's phrase, *Cardano's Cosmos*, p. 92.

14. BMSTR, 3:121.

15. Ibid., 3:121, 108.

16. Cardano, *Opera*, 5:32, 56.

17. The phrases are, respectively, adapted from Cesare Lombroso, as quoted by August Buck in his introduction to Cardano's *Opera*, 1:9; and from Grafton, *Cardano's Cosmos*, p. 198.

10. SOMETHING ABOUT AN EVIL SPIRIT

1. *Urkundenbuch der Universität Leipzig von 1409–1555*, edited by Bruno Stübel (Leipzig, 1879), p. 592 (no. 467).

2. Erler, *Die Matrikel*, p. 696.

3. Hermann Kesten, *Copernicus and His World* (London: Secker & Warburg, 1946), p. 304.

4. From the poem "Iter Rheticum" ("Rhaetian Journey"); reprinted in Ueli Dill and Beat R. Jenny, *Aus der Werkstatt der Amerbach-Edition* (Basel: Schwabe, 2000), pp. 252–256.

5. The letter appears in Adalbert Horawitz, *Caspar Bruschius: Ein Beitrag zur Geschichte des Humanismus und der Reformation* (Prague and Vienna, 1874), pp. 212–214. Latin text reprinted (and translated into German) in BMSTR, 3:73–76. The August 1547 dating of the letter is according to Jenny, *Aus der Werkstatt der Amerbach-Edition*, p. 129, note 113. For a full English translation, see the appendix, item 5.

6. Horawitz, *Caspar Bruschius*, p. 214.

7. *Briefwechsel der Brüder Ambrosius und Thomas Blaurer, 1509–1548*, edited by Traugott Schiess (Freiburg: Fehsenfeld, 1910), 2:674–675.

8. The letter is reprinted in Burmeister, "Rheticus—ein Bregenzer?" p. 315.

9. *Blaurer Briefwechsel*, 2:674.

10. *Urkundenbuch*, edited by Stübel, pp. 595–596 (no. 469).

11. The text appeared in Conrad Gesner, *Pandectarum sive partitionum universalium* (Zurich, 1548), part 1, pp. 80–81; reprinted in BMSTR, 3:77–80.

12. *Blaurer Briefwechsel*, 2:686 (Constance, February 18, 1548).

13. *Urkundenbuch,* edited by Stübel, pp. 602–603 (no. 476); reprinted in Burmeister, "Rheticus—ein Bregenzer?" p. 316.

11. Labors of Harvest

1. MCTHN, no. 4700.

2. Johannes Virdung, *Tabulae resolutae de supputandis siderum motibus* (Nuremberg, 1542), fol. 8.

3. Original manuscript in Muzeum Narodowe, Kraków; reprinted in BMSTR, 3:86.

4. *ACTA,* pp. 370–371; see also BMSTR, 1:103–104.

5. BMSTR, 3:90, 91.

6. Ibid., 3:95.

7. PROWE, 2:391.

8. The text of the charge (in German) appears in *ACTA,* pp. 394–395; reprinted in BMSTR, 1:110–111.

12. Things Left Behind

1. The account of the affair appears in intermittent entries in *ACTA* as follows: pp. 394–399, 406–407, 414–415, 419.

2. *Peinliche Gerichtsordnung,* p. 78. Article 116 reads: "Item so eyn mensch mit eynem vihe, mann mit mann, weib mit weib, vnkeusch treiben, die haben auch das leben verwürckt, vnd man soll sie der gemeynen gewonheyt nach mit dem fewer vom leben zum todt richten." For background on this topic, see Helmut Puff, *Sodomy in Reformation Germany and Switzerland, 1400–1600* (Chicago: University of Chicago Press, 2003), especially part 1.

3. BMSTR, 3:113.

4. Ibid.; "vel per procuratorem idoneum, sufficienter instructem."

5. BMSTR, 3:117–118.

6. *ACTA,* p. 419.

7. BMSTR, 1:118.

8. Karl Heinz Burmeister, "Ein Inventar des Georg Joachim Rhetikus von 1551," *Montfort* 56.3 (2004): 160–168.

9. Ibid., pp. 166–167.

10. *Wie man sich christlich zu dem sterben beraytten sol;* reprinted in Johannes Brenz, *Frühschriften,* part 2, edited by Martin Brecht et al. (Tübingen: Mohr/Siebeck, 1974), pp. 67–79.

11. BMSTR, 1:118; *ACTA,* p. 415.

12. Brenz, *Frühschriften,* p. 79.

13. MEDICINE IS NOT LIKE GEOMETRY

1. BMSTR, 1:125.

2. Ibid., 3:159.

3. Fabricius, dedication to *Oratio et Carmen de Carolo Quinto Caesare mortuo . . . Descriptio cometae* (Vienna, 1558); reprinted in BMSTR, 3:128–130. For a longer extract, in English translation, see the appendix, item 7.

4. BMSTR, 3:119.

5. Ibid., 3:121.

6. Ibid., 3:123.

7. Ibid., 3:139.

8. Ibid., 3:125–127.

9. Joseph F. Borzelleca, "Paracelsus: Herald of Modern Toxicology," *Toxicological Sciences* 53 (2000): 2–4.

10. Rheticus to Joachim Camerarius Jr., May 29, 1569; BMSTR, 3: 190–191.

11. Karl Heinz Burmeister, "Georg Joachim Rhetikus als Paracelsist," *Montfort* (1970):623.

12. Quoted in Allen G. Debus, "Paracelsus and the Medical Revolution of the Renaissance: A 500th Anniversary," in *Paracelsus, Five Hundred Years: Three American Exhibits* (Washington, DC: National Library of Medicine, 1999); URL: http://www.nlm.nih.gov/exhibition/paracelsus/paracelsus_2.html. The standard account is Walter Pagel, *Paracelsus: An Introduction to Philosophical Medicine in the Era of the Renaissance,* 2nd revised ed. (Basel: Karger, 1982).

13. Owen Gingerich, *The Book Nobody Read* (New York: Walker, 2004), p. 53.

14. Rheticus to Camerarius, November 9, 1558; BMSTR, 3:156.

15. BMSTR, 3:168.

14. ANOTHER COPERNICUS

1. The German lyrics of the hymn are as follows:

> *Wenn wir in höchsten Nöten sein*
> *Und wißen nicht, wo aus noch ein,*
> *Und finden weder Hilf' noch Rat,*
> *Ob wir gleich sorgen früh und spat:*
> *So ist dies unser Trost allein,*
> *Daß wir zusammen insgemein*
> *Dich rufen an, o treuer Gott,*
> *Um Rettung aus der Angst und Not.*

The hymn was based on Latin lyrics by Joachim Camerarius and was first published in the Genevan Psalter, 1547. The quoted English translation is by Catherine Winkworth, 1858.

2. BMSTR, 3:160. For the full letter in English translation, see the appendix, item 9.

3. See Avihu Zakai, "Reformation, History, and Eschatology in English Protestantism," *History and Theory* 26.3 (Oct. 1987): 303–304.

4. BMSTR, 3:162–163.

5. For more information on Rheticus's interest in Carion's *Chronicle*, see Catherine A. Tredwell's forthcoming essay "Melanchthon and Rheticus: Scripture, Cosmology, and History at Wittenberg."

6. BMSTR, 3:165. For the full letter of November 1562 in English translation, see the appendix, item 10.

7. BMSTR, 3:166–167.

8. Ibid., 3:171–172.

9. Ibid., 3:181.

10. BMSTR, 3:182–183.

11. "Cuperet te hic esse"; BMSTR, 3:169–170.

12. Ibid., 3:173–176.

13. See GINGR, pp. xv–xvi.

14. Lasicki's letters are printed in Theodor Wotschke, *Der Briefwechsel der Schweizer mit den Polen* (Leipzig: Heinsius, 1908), pp. 301–302. Excerpts are reproduced in Ludwik Antoni Birkenmajer, *Stromata Copernicana* (Kraków: Polish Academy, 1924), pp. 377–378.

15. Josiah Simler, *Bibliotheca Instituta et Collecta, primum a Conrado Gesnero* (1568; reprint, Zurich, 1583), p. 270; reprinted in BMSTR, 3: 187–188. For a translation of the full text, see the appendix, item 11.

16. *First Account*; Rosen*TCT*, p. 138.

17. Petrus Ramus, *Schools of Mathematics (Scholarum mathematicarum libri XXXI)* (Basel, 1569), p. 66: "ac nisi medicinam mecoenatis loco perdiscere et exercere coactus esset, jampridem alterum Copernicum mathemata celebrarent" (translation adapted from Edward Rosen, "The Ramus-Rheticus Correspondence," *Journal of the History of Ideas* 1.3 [June 1940]: 363–368).

15. Rescuing Rheticus

1. The foremost discussion of these "domains" is Westman, "The Astronomer's Role," 105–147.

2. Ramus, *Scholarum mathematicarum libri XXXI*, p. 66.

3. BMSTR, 3:186.

4. The two indispensable articles on Rheticus's relation to Paracelsian literature are by Karl Heinz Burmeister: "Georg Joachim Rhetikus als Paracelsist," *Montfort* (1972): 619–629; and "Die chemischen Schriften des Georg Joachim Rhetikus," *Organon* 10 (1974): 177–185.

5. Cited in Burmeister, "Die chemischen Schriften," p. 184. It is ironic that after his death, when cited as a source of knowledge about medicinal chemistry—a pursuit that almost thwarted the fulfillment of his mathematical destiny—Rheticus would still be referred to as "the renowned mathematician."

6. Gesner to Crato, in Gesner, *Epistolarum Medicinalium* (Zurich, 1577), fol. 2r; quoted in Ralf Bröer, "Friedenspolitik durch Verketzerung: Johannes Crato (1519–1585) und die Denunziation der Paracelsisten als Arianer," *Medizinhistorisches Journal* 37 (2002): 160–161.

7. Crato, *Methodus Therapeutica* (Basel, 1563), fol. 6v–7r; quoted in Bröer, "Friedenspolitik durch Verketzerung," pp. 140–141.

8. Quoted by Burmeister, *Achilles Pirmin Gasser*, 1:139.

9. Pagel, *Paracelsus*, p. 313.

10. Łukasz Górnicki, *Der polnische Demokrit als Hofmann* (Stuttgart, 1856), pp. 247–248. Reference kindly provided by Karl Heinz Burmeister in private correspondence.

11. Joachim Camerarius Sr. to Dudith, Leipzig, March 29, 1569; DUDITH, 2:88–89.

12. Dudith to Camerarius, Kraków, May 15, 1569; ibid., 2:103.

13. Dudith to Camerarius, Kraków, February 8, 1570; ibid., 2:123.

14. Caspar Peucer, *Hypotheses astronomicae, seu theoriae planetarum ex Ptolemaei et aliorum veterum doctrina ad observationes Nicolai Copernici accommodatae* (Wittenberg, 1571), the preface.

15. On Dasypodius and the Strasbourg clock, see Günther Oestmann, *Die astronomische Uhr des Strassburger Münsters: Funktion und Bedeutung eines Kosmos-Modells des 16. Jahrhunderts* (Stuttgart: Verlag für Geschichte der Naturwissenschaften und der Technik, 1993).

16. For some of Dasypodius's other projects under way at this time, see Abraham Gotthelf Kästner, *Geschichte der Mathematik*, vol. 1 (1796; reprint, Hildesheim and New York: Georg Olms, 1970), pp. 332–345.

17. Dasypodius to Dudith, December 3, 1571; DUDITH, 2:322. Dasypodius had undertaken a massive program of collecting and editing ancient mathematical manuscripts. See Oestmann, *Die astronomische Uhr*, p. 38.

18. See *De nova stella*, translated by John H. Walden, *A Source Book in Astronomy*, edited by Harlow Shapley and Helen E. Howarth (New York: McGraw-Hill, 1929); reprinted in *BOTC*, p. 129.

19. See Charlotte Methuen, " 'This Comet or New Star': Theology and the Interpretation of the Nova of 1572," *Perspectives on Science* 5.4 (1997): 499–515.

20. Maximilian to Dudith, Nov. 12, 1573; DUDITH, 2:539.

21. Dudith to Hájek, Kraków, April 12, 1573; ibid., 2:395. In Hungary Rheticus likely continued to work on his translations of Paracelsus. Michael Toxites reported in 1574 that Rheticus's edition of the *Archidoxa* was just about to appear. See Burmeister, "Die chemischen Schriften," p. 183.

22. See Menso Folkerts, "Johannes Praetorius (1537–1616)—ein bedeutender Mathematiker und Astronom des 16. Jahrhunderts," in *History of Mathematics: States of the Art*, edited by Joseph W. Dauben et al. (San Diego: Academic Press, 1996), pp. 152, 163. Praetorius also owned and annotated copies of the very earliest works by both Rheticus and Copernicus. See GINGR, pp. 91–93, 308–313; and Robert S. Westman, "Three Responses to the Copernican Theory: Johannes Praetorius, Tycho Brahe, and Michael Maestlin," in *The Copernican Achievement*, edited by Westman, pp. 289–305.

23. All quotations from Otto are from his preface to the reader, *Opus Palatinum*. In Latin the last two sentences read: "Ego his ita excitor et incen-

dior, ut temperare mihi non possim, quin primo quoque tempore ipsum autorem adeam et coram de singulis cognoscam. Profectus itaque in Ungariam, ubi tum agebat Rheticus."

24. A. von Braunmühl writes in his history of trigonometry (*Vorlesungen über Geschichte der Trigonometrie* [Leipzig: Teubner, 1900]) that the appearance of Rheticus's gigantic work on triangles "was looked forward to with longing by all astronomers" ("[wurde] von allen Astronomen mit Sehnsucht erwartet") (1:212). Rheticus's contacts included Ramus in Paris, Gesner in Zurich, Hájek in Prague, Crato and Fabricius in Vienna, Camerarius in Leipzig, Dudith, Schuler, and Praetorius in Kraków, and Peucer in Wittenberg. By 1574 Gesner and Ramus had died, and Schuler and Praetorious had moved from Kraków and taken up their professorships in Wittenberg, where Otto was one of their students. Of these men, Ramus, Fabricius, Dudith, Schuler, Praetorius, and Peucer had at various times either personally or publicly expressed their desire that Rheticus should devote himself more fully to completing his mathematical projects. A number of them—Crato, Fabricius, and Dudith—were in direct contact with Maximilian II. Johann Rueber, who had invited Rheticus to that region in the first place, was a correspondent of Dudith's and a Protestant known for sending students all the way to Wittenberg for their education (see DUDITH, 2:436).

16. TRIANGLES, STARS, AND THE SWEETNESS OF THINGS

1. Rheticus's letter to King Ferdinand; see the appendix, item 8.

2. All quotations from Otto are taken from his preface to the reader, *Opus palatinum.*

3. Adrianus Romanus, *Ideae mathematicae*, part 1 (Antwerp, 1593), "Lectori philomathi," sig. **iir.

4. Manuscript letter from Romanus to Johann Georg Herwart von Hohenburg, Munich Universitätsbibliothek, manuscript 2° 692, pp. 126–129; quoted and translated in Paul Bockstaele, "Adrianus Romanus and the Trigonometric Tables of Georg Joachim Rheticus," in *Amphora: Festschrift for Hans Wussing on the Occasion of his 65th Birthday*, edited by S. Demidov et al. (Basel: Birkhäuser Verlag, 1992), pp. 60–61.

5. BMSTR, 1:180.

6. In a July 1600 letter Kepler refers to "Otto, author of that *Opus palatinum*, who is presently staying in Prague"; Kepler*GW*, 14:131.

7. Pitiscus, *Thesaurus mathematicus* (Frankfurt, 1613), "Ad lectorem." For a fuller account of Pitiscus's contribution, see Kästner, *Geschichte der Mathematik*, 1:612–626, and von Braunmühl, *Vorlesungen über Geschichte der Trigonometrie*, 1:220–226.

8. R. C. Archibald, "Rheticus, with Special Reference to his *Opus Palatinum*," *Mathematical Tables and Other Aids to Computation*, 3.28 (Oct. 1949): 557, 558. See also the note "Bartholomäus Pitiscus" by the same author and in the same journal, 2.25 (Jan. 1949): 390–397.

9. Glen van Brummelen, who is preparing a new history of trigonometry; personal correspondence, Sept. 2005.

10. Georg Ludwig Froben, *Clavis universi trigonometrica* (Hamburg, 1634).

11. Edward Sherburne, *The Sphere of Marcus Manilius* (London, 1675), appendix, p. 55.

12. See GINGR, pp. 114, 119–120, 125, 131, 157–158, 165, 169, 188, 200–201, 331, 341, 346.

13. See the diagram placed between pages 26 and 27 of Kepler*GW*, vol. 1.

14. Rosen*TCT*, p. 147.

15. Max Caspar, Kepler*GW*, 1:418.

16. Rosen*TCT*, p. 168.

17. Kepler*GW*, 1:116. See Katherine A. Tredwell's excellent discussion, "Michael Maestlin and the Fate of the *Narratio Prima*," *Journal for the History of Astronomy* 35 (2004): 305–325.

18. Hans Blumenberg, *Die Genesis der kopernikanischen Welt* (Frankfurt: Suhrkamkp, 1975), p. 396. See Blumenberg's entire chapter (part 4, chap. 4), even though its view of Rheticus (whom Blumenberg thinks "not far from being a [Copernican] apostate") differs in important respects from that of this book. Blumenberg's book was also published as *The Genesis of the Copernican World*, translated by Robert M. Wallace (Cambridge, MA: MIT Press, 1987).

APPENDIX: TRANSLATIONS OF SOME DOCUMENTS RELATING TO THE CAREER OF RHETICUS

1. Hipler, *Spicilegium Copernicanum*, pp. 351–352; quoted in BMSTR, 1:47.

2. BMSTR, 3:15–18.

3. Reprinted in " 'Mit subtilen fündlein . . . '," pp. 107–120.

4. Reprinted in ibid.

5. Horawitz, *Caspar Bruschius,* pp. 212–214; reprinted in BMSTR, 3: 73–76.

6. The original does not offer a breakdown by category. For an exact transcription of the inventory together with brilliant detailed commentary on it, see Burmeister, "Ein Inventar . . . ," pp. 160–168.

7. From Fabricius's dedication to *Oratio et carmen;* reprinted in BMSTR, 3:128–130.

8. From Rheticus's edition of *Ioannis Verneri De triangulis sphaericis libri quatuor* [and] *De meteoroscopiis libri sex* (Kraków, 1557); reprinted in BMSTR, 3:132–140.

9. BMSTR, 3:160.

10. Ibid., 3:165.

11. Published in Simler, *Bibliotheca,* p. 270; reprinted in BMSTR, 3:187–188.

INDEX

NOTE: GJR refers to Rheticus.

A NOTE ON THE AUTHOR

Dennis Danielson is a professor of English at the University of British Columbia. He has served as a member of the History-of-Astronomy Committee for the Adler Planetarium and Astronomy Museum in Chicago, and is a member of the Historical Astronomy division of the American Astronomical Society. His articles have appeared in the *American Journal of Physics* and the *Journal for the History of Astronomy*. He is the editor of the acclaimed anthology of cosmological writings *The Book of the Cosmos*. He lives in Vancouver, British Columbia.